Schriftenreihe des Energie-Forschungszentrums Niedersachsen (EFZN)

Band 66

Das EFZN ist ein gemeinsames
wissenschaftliches Zentrum der
Universitäten:

Development of coupled THM models for reservoir stimulation and geo-energy production with supercritical CO_2 as working fluid

Dissertation

to be awarded the Degree of

Doctor of Engineering (Dr.-Ing.)

submitted by

M.Sc. Jianxing Liao

from Guizhuo, P.R. China

approved by the Faculty of

Energy and Economic Sciences,

Clausthal University of Technology

Date of oral examination

June 30, 2020

Bibliografische Information der Deutschen Nationalbibliothek

Die Deutsche Nationalbibliothek verzeichnet diese Publikation in der Deutschen Nationalbibliografie; detaillierte bibliographische Daten sind im Internet über http://dnb.d-nb.de abrufbar.

1. Aufl. - Göttingen: Cuvillier, 2020

   Zugl.: (TU) Clausthal, Univ., Diss., 2020

D 104

Dean

Prof. Dr. rer. nat. habil. Bernd Lehmann

<in case:> Chairperson of the Board of Examiners

Supervising tutor

Prof. Dr.-Ing. habil. Michael Zhengmeng Hou

Reviewer

Prof. Dr.-Ing. habil. Olaf Kolditz

© CUVILLIER VERLAG, Göttingen 2020

   Nonnenstieg 8, 37075 Göttingen

   Telefon: 0551-54724-0

   Telefax: 0551-54724-21

   www.cuvillier.de

   ISBN 978-3-7369-7242-1

   eISBN 978-3-7369-6242-2

Acknowledgements

I would like to express my sincere thanks to all the people who kindly offered me help during my PhD study at the institute of petroleum engineering, Clausthal University of Technology, Germany.

First and foremost, I would like to express my gratitude to my supervisors Prof. Dr. -Ing. habil. Michael Zhengmeng Hou for his consistent support, motivation and guidance. I also would like to thank Prof. Dr. -Ing. habil. Olaf Kolditz for entrusting me with the task of preparing this thesis under their untiring guidance. It was a great pleasure for me to work and learn from them.

Next, I would like to express my sincere appreciation to the industrial partners in the DGMK-814-project as well as the scholarship (CSC201708080145) from china scholarship council.

My appreciation also extends to my colleagues at the institute of petroleum engineering and the research center energy storage technologies for a good sense of teamwork as well as wonderful time. Thanks also go to Faisal Mehmood, who devoted more time to carefully review and revise my thesis for its grammatical order. I would like to thank Dr. Yang Gou, for his great help in solving the technical problems related to the numerical simulations. Miss Peya's genuine kindness also helped sustain a positive atmosphere in which to do science.

Above ground, I am indebted to my family, whose value to me only grows with age. Last but not least, I would like to give my dearest thanks to my girl friend Zhengyi Chen. Without her continuous moral support and encouragement, I could not have been able to finish my thesis in the stipulated time.

Life becomes more beautiful after we overcome the difficulties one by one. After the continuous improvement, one can become the person whom one wants.

Jianxing Liao

Clausthal-Zellerfeld, April 2020

Abstract

Due to enhanced public awareness, the eco-friendly techniques in geo-energy exploitation, e.g. supercritical CO_2 fracturing, have received extensive attention in the last two decades. In this dissertation, two specific numerical models have been developed to address the issues associated with utilization of supercritical CO_2, like fracture creation, proppant placement and fracture closure in unconventional gas reservoirs, reservoir stimulation, heat production and CO_2 sequestration in deep geothermal reservoirs, respectively. (a) In unconventional gas reservoir, the model consisting of classic fracture model, proppant transport model as well as temperature-sensitive fracturing fluids (CO_2, thickened CO_2 and guar gum) has been integrated into the popular THM coupled framework (TOUGH2MP-FLAC3D), which has the ability to simulate single fracture propagation driven by different fracturing fluids in non-isothermal condition. (b) To characterize the fracture network propagation and internal multi fluids behavior in deep geothermal reservoirs, an anisotropic permeability model on the foundation of the continuum anisotropic damage model has been developed and integrated into the popular THM coupled framework (TOUGH2MP-FLAC3D) as well. This model has the potential to simulate the reservoir stimulation and heat extraction based on a CO2-EGS concept.

Attempting to improve the fracturing ability and proppant-carrying capacity of CO_2, the unconventional gas reservoir model is applied in a fictitious model with the parameters of a typical tight gas reservoir. According to results, pure CO_2 is inefficient to create a fracture in such tight gas reservoir. Its fracturing ability can be improved significantly by using CO_2 thickener. In comparison with traditional guar gum, the leakage of thickened CO_2 is higher. But considering the expansion of CO_2 in the fracture, accounting for about 37% contribution in this case, the performance of thickened CO_2 is comparable with traditional guar gum. Additionally, a linear correlation between the break down pressure and fracturing fluid viscosity has been observed. In non-isothermal conditions, the temporal and spatial leak-off rate and expansion effect distribute unevenly in fracture. Due to high temperature around fracture tip, High leak-off rate and expansion effect is found nearby the tip zone. During fracture initiation, the expansion effect plays a critical role. With continuous injection, the dominating factor in fracture-making is gradually switched to leak-off. Overall, fracturing is more sensitive to leak-off than fluid expansion. For proppant-carrying, thickened CO_2 with light proppant can achieve a better proppant placement than heavy proppant, even better than the one transported by guar gum. Due to low specific heat capacity, thickened CO_2 has a high gel breaking rate, resulting in better fracture support.

In the application of the damage-permeability model at planned Dikili geothermal project, a fracture network with final SRV of 8.7×10^7 m³ and SRA of 2.2×10^6 m² is created after injecting 90,000 kg CO_2 in 250 hours. The maximum permeability of 1.3×10^{-13} m² has been achieved in x- and z-directions. During heat production, a priority channel with high gas saturation is formed because of the viscosity difference between CO_2 and water. The heat is mainly mined in this priority channel. Generally, for both water and CO_2 injection, the driven pressure and average thermal capacity shows a positive correlation with injection rate, while inverse relation exists between eventual temperature and injection rate. Under the same mass injection rate, the driven pressure and average thermal capacity of water injection is higher than CO_2 injection, whereas the ultimate produced temperature of water injection is lower. Therefore, CO_2 as working fluid for heat extraction has the benefits of low driving pressure and is beneficial for realizing relatively stable heat mining. In addition, the injected CO_2 is detained in geothermal reservoir due to leak off, which can be regarded as geologically sequestrated CO_2. Furthermore, the sequestrated CO_2 has a positive relationship with injection rate. After 30 years of geothermal production, up to 950,000 tons CO_2 is sequestrated at an injection rate of 100kg/s, demonstrating the potential contribution of CO_2-ESG on the sequestration of CO_2.

List of contents

List of tables

1. Introduction

1.1. Motivation, objectives and scientific challenges

In the commercial exploitation of unconventional gas reservoir, for instance tight and shale gas reservoir, hydraulic fracturing plays a very important role. It is used to enhance the original permeability of unconventional reservoirs, which usually is very low even ultra-low. During hydraulic fracturing, the fracturing fluid, normally composed of water and some chemical additives, is pumped from surface down to perforation through injection well. The pressurized fluid breaks the rock in tension firstly surrounding perforations, and propagates into deep reservoir. To prevent the closure of opened fracture after removing pumping power, some proppant is injected along with fracturing fluid, when enough fracture has been created. The created artificial fracture performs as a high conductive channel connecting tight reservoir and production well for gas flow, which could enhance the gas production significantly in comparison with original reservoir. Thanks to the innovation of horizontal and directional drilling technology, currently, a horizontal well can be drilled into the reservoir over 1000m (Ma et al. 2016). As shown in Fig. 1.1, the horizontal well will be divided into 10 or more segments for fracturing. In this case, multi-fractures can be stimulated through only one well, which greatly improve production efficiency and makes the exploiting of unconventional reservoir more economic.

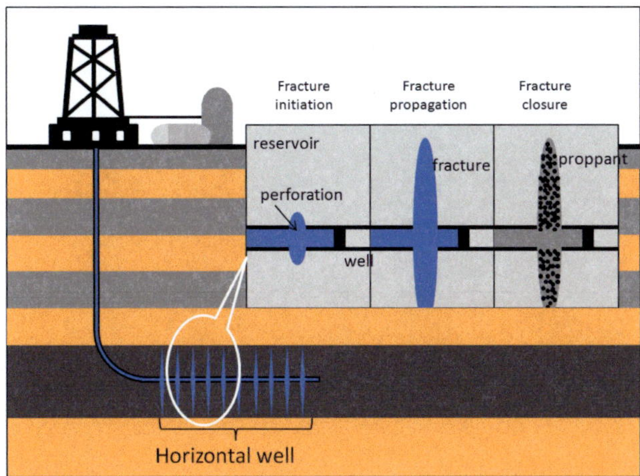

Figure 1.1. Demonstration of hydraulic fracturing in gas reservoir

Geothermal energy is kinds of thermal energy generated and stored inside the earth. Due to its excellent features like huge capacity and independency on weather conditions, geothermal energy as one of the important renewable resources attracts more and more attentions, especially hot dry rock (HDR), which is impervious crystalline basement rock with remarkable temperature (Armstead & Tester 1978). The HDR owning a vast capacity can be found almost everywhere deep beneath the earth's surface. But, extraction and utilization of this heat resource remains an enormous challenge. Enhanced geothermal system (EGS) technology has a goal to extract the heat out of HDR for electricity generating. As shown in Fig.1.2, the EGS is composed of at least two well, one for injection and one for production. The cold medium, usually pure water is pumped down to target geothermal reservoir, where the cold medium will be heated through energy exchange with HDR. Then the heated medium will be brought up to surface again for electricity generating. Such circulation makes the utilization of HDR resource possible. However, before circulation, the permeability of geothermal reservoir needs been enhanced for efficient heat mining, generally by using hydraulic fracturing. Like in unconventional reservoir, the hydraulic fracturing plays a very important role in EGS, and is used to create artificial fractures for improving the geothermal reservoir conductivity.

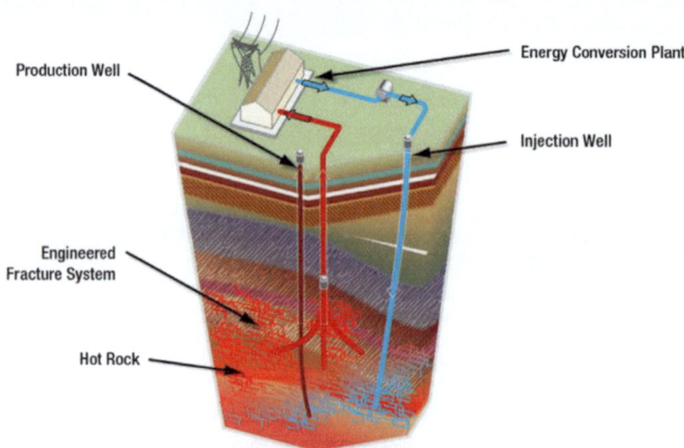

Figure 1.2. Demonstration of an enhanced geothermal system (Olasolo et al. 2016)

The reservoir stimulation in unconventional gas reservoir shares some common characteristics with the one in HDR reservoir. For the inefficient conductivity, both reservoirs need be stimulated by pressurized fracturing fluid to enhance the original permeability before production. And the hydraulic fracturing process is a thermal-hydraulic-mechanic (THM) coupled process, involving heat flow, multi-component and -phase flow, mechanic response of reservoir rock. Besides, massively implementation of water-based fracturing in unconventional gas reservoir as well as DHR could induce water shortage or conflicts with other users, particularly in water-scare area (Vengosh et al. 2014). In addition to this, there are also some differences in reservoir stimulation and production. Firstly, the host rock type is different. Unconventional gas reservoir is found almost in porous sedimentary, whereas the HDR is consisted mostly of crystalline volcanic rock. Normally a signal fracture with two wings is stimulated in unconventional gas reservoir, if no natural fracture is contained in sedimentary. Due to high-density of naturally fractures, usually a complex fracture network is stimulated through hydraulic fracturing in volcanic rock. And the temperature in HDR often over 160°C (Lu 2018) is higher than in unconventional gas reservoir. Additionally, proppant is not necessary in reservoir stimulation of HDR, because during heat production, the injection fluid provides the pressure to support fracture wall.

Due to the low cost, availability and high specific heat, currently water is the mostly used working fluid for unconventional gas reservoir stimulation as well as HDR reservoir stimulation. In general, some chemical additives are added into fracturing fluid to improve its performance on fracture making ability and proppant carrying capacity. Hence, water-based fracturing often brings some environmental problem, like groundwater pollution and formation damage, water shortage, unsuitable in water-sensitive clay mineral (Ward et al. 2016, Schill et al. 2017, Sanaei et al. 2018). To address these problems, some alternative non-aqueous fracturing fluids have been proposed. Several important non-aqueous fracturing fluids, including foam-, oil-, alcohol-, Emulsion-based fracturing fluid, and liquid CO_2 and N_2, are summarized in Tab.1.1. In addition to the common characteristics, liquid CO_2 shows some unique advantages, including CO_2 sequestration, enhancing gas recovery by displacing the adsorbed methane, rapid clean-up etc. The CO_2 will presents as supercritical phase, if the temperature and pressure exceed 31.04°C and 71.82 bars at the same time (Vargaftik 1975). This condition could be easily met under fracturing in unconventional gas reservoir. For this reason, the CO_2 in fracture exists mostly as supercritical CO_2 during fracturing. Not only in unconventional gas reservoir, but in EGS the CO_2 can be used as working fluid to exact geothermal energy. Apparently, the CO_2 running EGS circulation is supercritical. However, supercritical CO_2 stills a concept in reservoir stimulation and geothermal energy production. Before commercial application of supercritical CO_2, some barriers, e.g. proppant transport, corrosive of CO_2 dissolution, transport etc. need be removed.

To gain deeper understanding on the CO_2 behavior in geo-energy exploitation, numerical methods on the base of THM coupled framework will be used to study CO_2-based hydraulic fracturing in tight gas reservoir, CO_2-based reservoir stimulation and heat extraction in HDR reservoir, respectively.

Table 1.1. Summary of potential advantages and disadvantages of several non-aqueous fracturing fluids (adapted from Gandossi 2013)

Fracturing fluids	Potential advantages	Potential disadvantages	Status of applications
Foam-based	-Water usage reduced -Reduced amount of chemical additives. -Reduction of formation damage. -Better clean-up of the residual fluid	-Low proppant concentration in fluid, -Higher costs. -Difficult rheological characterization -Higher surface pumping pressure required.	Some commercial applications
Oil-based (LPG)	-Water usage much reduced or completely eliminated. -Fewer chemical additives are required. -Flaring reduced. -Truck traffic reduced. -Abundant by-product of the natural gas industry. Increased the productivity of the well. -Excellent fluid properties -Full fluid compatibility with shale reservoirs -No fluid loss. -Recovery rates (up to 100%) possible. -Very rapid clean up	-Involves the manipulation of large amounts of flammable propane, hence potentially riskier than other fluids and more suitable in environments with low population density. -Higher investment costs. -Success relies on the formation ability to return most of the propane back to surface to reduce the overall cost.	Some commercial applications in unconventional reservoir
Alcohol-based (Methanol)	-Water usage much reduced or completely eliminated. -Methanol is not persistent in the environment. -Excellent fluid properties -Very good fluid for water-sensitive formations.	-Methanol is a dangerous substance to handle: a. Low flash point, hence easier to ignite. b. Large range of explosive limits. c. High vapour density. d. Invisibility of the flame.	It has been used in low permeability reservoir
Emulsion based	-Depending on the type of components used to formulate the emulsion, these fluids can have potential advantages such as: a. Water usage much reduced or completely eliminated. b. Fewer (or no) chemical additives are required. -Increased the productivity of the well; better rheological properties	-Potentially higher costs.	It has been used on unconventional reservoir
Liquid CO_2	-Potential environmental advantages: a. Water usage much reduced or completely eliminated. -Fewer chemical additives -Some level of CO_2 sequestration achieved. -Reduction of formation damage. -Form more complex micro-fractures. -Enhance gas recovery by displacing the methane adsorbed in the shale formations. -Rapid clean-up. -More controlled proppant placement and higher proppant placement within the created fracture width.	-Proppant concentration must necessarily be lower and proppant sizes smaller, hence decreased fracture conductivity. -CO_2 must be transported and stored under pressure (typically 2 MPa, -30°C). -Corrosive nature of CO_2 in presence of H_2O -Unclear (potentially high) treatment costs.	Liquid CO_2 as fracturing fluid is commercially used in unconventional applications Supercritical CO_2 usage appears to be at the concept stage.
Liquid N_2	-Water usage much reduced or completely eliminated. -Fewer chemical additives are required. -Reduction of formation damage. -Self-propping fractures can be created by the thermal shock, hence need for proppant reduced or eliminated	-Can replace hydraulic fracturing only for small to medium treatments -Proppant is not carried into the fracture. Instead, propellant fracturing relies upon shear slippage to prevent the fracture from fully closing back on itself. -Potentially induce seismic events.	The use of liquid nitrogen is less typical

Because of a relatively new concept, CO_2 for reservoir stimulation and production is still in the preliminary study stage. Rare applications has been conducted and planned in commercial projects. In spite of this, there remain several important numerical models have been proposed to simulate CO_2 fracturing in unconventional gas reservoir. Jiang et al. (2014) developed a multi-Continuum multi-component model to study the CO_2 performance on enhanced gas recovery in a shale gas reservoir with complex fracture network. In this reservoir, different structures, involving matrix, major hydraulic fracture and large-scale natural fractures, were described by embedded discrete fracture model. Lee and Ni (2015) analyzed CO_2 behavior in discrete fracture network, wherein the flow and CO_2 migration was simulated by using flow code THOUGH2/ECO2N. The TOUGH series code is primarily developed at Lawrence Berkeley National Laboratory (LBNL) (Pruess 2004) and as one of the popular multi-component and -phase programs has been widely used to study the multi-flow in unconventional gas reservoir. Fang et al. (2014) used a 2D commercial program, universal distinct element code (UDEC) to compare the fracture-making ability of different fluids: water, slick water and CO_2. The above-mentioned models focus only on the flow and CO_2 behavior in predefined fracture or complex fracture network. Yet, the fracture initiation and propagation have not been considered. Then, Peng et al. (2017) conducted the simulation on supercritical CO_2 fracturing by using bonded particle model, in which the rock material was modeled as a collection of round particles and the microcracks was modeled by breakage of bound between particles. The bonded particle model performs very well in characterizing fracture initiation and propagation on micro level, but for large scale project it is still inefficient. More recently, the 2D damaged-based model (Liu et al. 2018, Wang et al. 2018, Zhang et al. 2019) was proposed to investigate supercritical CO_2 fracturing. In this model, the fracture pattern was characterized by the damage induced by rock-gas interaction due to CO_2 injection. During fracture propagation, equivalent permeability of corresponding fracture was updated according to value of damage. However, this model is still limit to solve the fracture field scale fracture propagation and proppant transport. Comparing with unconventional gas reservoir, the study on CO_2 as working fluid to exploit geothermal energy starts relatively late. Since Brown (2000) firstly proposed the CO_2-EGS concept, some preliminarily works has been conducted. The numerical simulator for CO_2-EGS could be principally divided into three categories. One focused only on supercritical CO_2 performance in operation of naturally permeable or already stimulated geothermal reservoir, in which the reservoir stimulation and gas-interaction was ignored (Pruess and Azaroual 2006, Pruess 207, Randolph and Saar 2010, Luo et al. 2014, Shi et al. 2018). The gas flow and heat exchange efficiency of supercritical CO_2 EGS have been discussed in these models. Because the CO_2 dissolving in water exhibits acids, the CO_2 dissolution can chemically react with surrounding minerals, further causing mineral dissolution or precipitation, which has potential influence on the productiveness of geothermal reservoir. Thus another category of simulators was proposed to interpret the effects of chemical interaction on EGS performance (Xu and Pruess 2010, Wan et al. 2011, Borgia et al. 2012, Cui et al. 2018). Additionally, mechanical interaction of hot rock plays also a very important role in heat extraction. In the third category, the mechanical response of hot rock has been considered during CO_2 injecting into a discrete fracture network, in which the cracks width was local pressure- and stress-dependent (Sun et al. 2017, Li et al. 2019a). However, the reservoir stimulation before production has not been considered in these models.

At present, the primary numerical method for hydraulic fracturing includes finite element method (FEM), distinct element method (DEM), finite difference method/finite volume method (FDM/FVM) and boundary element method (BEM). They have their own advantages and disadvantages. FEM is a very popular method in solid mechanic, because of the complete theory. On base of EFM, the extended finite element method (XFEM) is developed mainly to solve the fracture propagation, which can be used to simulate hydraulic fracturing (Zhao et al. 2014, Wang 2015, Zhou et al. 2015,Yan et al. 2016). In this method, an enrichment function is introduced to describe continue and discontinue deformation around fracture as well as fracture tip. As a consequence, re-mesh at fracture tip is not necessary and improve the competing efficient. However, XEFM is currently not suitable for 3D problem. In DEM, the pore rock material is represented by a collection of micro particles (Al-Busaidi et al. 2005, Yoon et al. 2015, Damjanac and Cundal 2016). The mechanic response between micro particles is transferred by the assumed springs in both normal and tangential directions, which performs as tensile/compress and shear strength on macro level. The fluid can flow in the pores between particles even break the assumed springs, to create a hydraulic fracture. DEM is very visible method to simulate shear and tensile fracture. But it is not efficient for field scale simulation if small particles are chosen. FDM/FVM is often used to solve fluid as well as heat flow (Wei and Zhang 2013, Wang et al. 2014). It is one of the most popular methods in simulation of reservoir flow, due to its complete theory and well development. In this method, the stress and pressure of element is represented by the one averaged on the center point of control volume, rather than from the grid point. Therefore, the converted scalar stress loses much information about direction. On the other side, it is also limit to solve the propagation direction, as the fracture parts is pre-defined during model generation. The basic idea of BEM is converting area integration to curve integration, or converting volume integration to area integration with the help of Green's function (Olson 2004, Olson 2009, Wu and Olson 2013, Wu and Olson 2015). Then the whole problem is converted to boundary problem. Through integration on discrete boundary can achieve the stress as well as deformation distribution inter domain. By treating the fracture wall as inter discrete boundary, this method could be used to simulate hydraulic fracturing. Dimension reduction improves the computing efficient significantly. Despite, the precondition, homogenous mechanic properties within the domain, limits its application.

The main scientific challenge of this thesis is development of suitable numerical models to study the mechanic and hydraulic behavior in utilization of supercritical CO_2 in corresponding geo-energy production. More concretely, develop THM coupled model to characterize fracture network growing during hydraulic fracturing in deep geothermal reservoir; study the reservoir stimulation, heat exaction and CO_2 sequestration in deep geothermal reservoir with pre-existing formation water based on CO_2-EGS concept; develop THM coupled model for hydraulic fracturing in unconventional gas reservoir, involving at least fracture model, proppant transport model as well as temperature sensitive fracturing fluid (CO_2, thickened CO_2 and guar gum); improve the fracture-making ability and proppant-carrying capacity of CO_2 in unconventional gas reservoir; compare the performance of supercritical CO_2 with traditional working fluid, like water and guar gum in geo-energy production.

1.2. Thesis outline

In this work, the models are developed on the fundament of popular THM framework TOUGH2MP-FLAC3D, in which both codes are developed on the base of FDM. In order to overcome its shortness for information loss, iso-parameter method is applied to interpret the local stress state. As Fig. 1.3 shown, different specific models are integrated in this THM framework, to study the CO_2 performance in different reservoir, including HDR reservoir and unconventional gas reservoir. Generally, in HDR reservoir, especially in volcanic rock types, usually a fracture network is stimulated by hydraulic fracturing. Therefore, an anisotropic continuum damage-permeability model is introduced into describe the equivalent anisotropic permeability as well as porosity for stimulated HDR reservoir. The feasibility of such models have been verified by measured stress, damage, permeability in series triaxial tests as well as a small scale field testing in hard rock. Then the developed model is applied in a specific planned Dikili EGS to study the performance of CO_2 in reservoir stimulation and heat extraction, comparing with water. To solve the hydraulic fracturing in unconventional gas reservoir, another model, containing fracture model (single fracture model), proppant model (proppant transport and placement), and fracturing fluid model (water, guar gum, pure CO_2 and thickened CO_2) has been integrated into the THM framework and verified, separately. In these model, the fracture initiation, propagation, proppant transport and placement, fracture closure even gel break can be simulated based on different working fluids in unconventional gas reservoir, because the temperature-sensitive properties of fracturing fluid, like viscosity, density, enthalpy etc. has been integrated in these models. Then the performance of different fluid, thermal effect as well as proppant transport has been compared using the newly developed models.

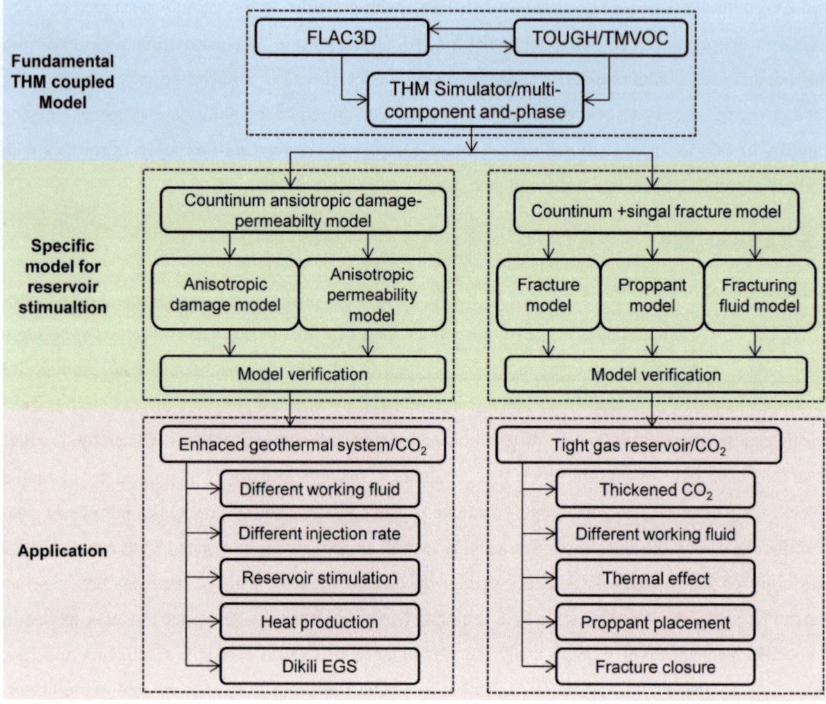

Figure 1.3. Research content and flowchart

The content of this thesis is divided into five chapters. In chapter 2, the state-of-the-art of CO_2 engineered application in geo-production will be studied through literature study. The advantages and disadvantages will be compared systematically. Besides, it is trying to point the critical barrier for industrial application and the potential solutions.

As the property of CO_2 is very temperature and pressure sensitive and many complex mechanic responses are involved in energy production, the new developed models should be integrated under a THM coupled framework. Therefore, the fundamentals of used THM framework (TOUGH2MP-FLAC3D), including theoretical background, governing equation, numerical algorithm as well coupling process, are introduced in chapter 3 in detail.

Attempting to learn the behavior of supercritical CO_2 in unconventional gas reservoir, a specific numerical model able to characterize fracture propagation and leak-off will be developed in chapter 4. In this model, a pre-defined potential fracture plane is inserted in the middle of host reservoir elements perpendicular to minimum stress. On this potential fracture plane, different specific models, including fracture model, proppant transport model and multi flow model, are integrated in the popular THM framework TOUGH2MP-FLAC3D. Then, the feasibility of these implemented models will be verified separately by analytical solution, experimental results and other simulated results. In this chapter, different fracturing fluids (mainly temperature-dependent viscosity property) is defined firstly, including pure CO_2, different thickened CO_2s and guar gum. Other properties, like density and enthalpy, is assumed equal to the base fluid of fracturing fluid. Then, different fracturing fluids, pure CO_2, two thickened CO_2s, pure water and guar gum, will be injected into a fictitious model with the properties of a typical tight gas reservoir, to compare their performances in fracture making-ability. Additionally, since the property of CO_2 is closely correlated to temperature, the fracturing by using CO_2 will be carried out under different thermal condition. Finally, proppant with different density is used to address the weakly proppant-carrying ability of CO_2.

In chapter 5, in order to study the critical engineering process, reservoir stimulation and heat extraction in deep geothermal reservoirs, a 3D simulator based on a continuum anisotropic model and anisotropic permeability model will be developed under the THM framework of FLAC3D-TOUGH (TMVOC). In geothermal energy exploitation, the stress state change due to fracturing fluid injection could induce an anisotropic damage in geothermal reservoir, which further triggers an anisotropic permeability. Macroscopically, it performs as fracture propagation. In addition to this, the impact of effect stress will be considered in this model as well. It is believed that usually a fracture network is stimulated in the geothermal reservoir with brittle rocks, especially volcanic rock. This damage model and permeability will be verified by triaxial test in laboratory and cyclic fracturing test in flied. The influence factor will be discussed by a series of case studies on a 2D simple model. On a real case from Dikili geothermal project, the reservoir stimulation and heat extraction with different productions rates by using supercritical CO_2 will be conducted and compared with the results by using water, in which pre-existing formation water is taken considered. Besides, the ancillary benefit of CO_2 sequestration will be discussed in the lifetime of CO_2-EGS under different injection rate.

2. Engineered applications of CO$_2$ in geo-energy

2.1. The carbon dioxide challenge in climate change

Carbon dioxide (CO$_2$) is a chemical compound composed of one carbon and two oxygen atoms. It is an important composition of the earth's atmosphere, accounting for about 0.03%. This CO$_2$ has two important functions. One is the resource for plant photosynthesis, to convert the light power to bio power, at the same time releasing oxygen gas. Another one is the greenhouse gas to keep earth's temperature in a suitable range, with a minor day-night difference. However, large amount of CO$_2$ in atmosphere will cause global warming, further bring some worse problem, like more extreme weather events, species extinction, sea level rise etc. As shown in Fig. 2.1, the global averaged temperature is closely correlated to the CO$_2$ concentration in atmosphere. Along with the CO$_2$ concentration increasing from 280ppm in 1850 (pre-industrial) up to 408ppm in 2018, global averaged temperature has warmed about 1.2°C. The global warming trend will continue, if nothing is done in the future. Therefore, the goal to hold the global average warming within 2°C comparing with pre-industrial level is set in the Pairs Agreement and adopted by many nations (IPCC, 2018).

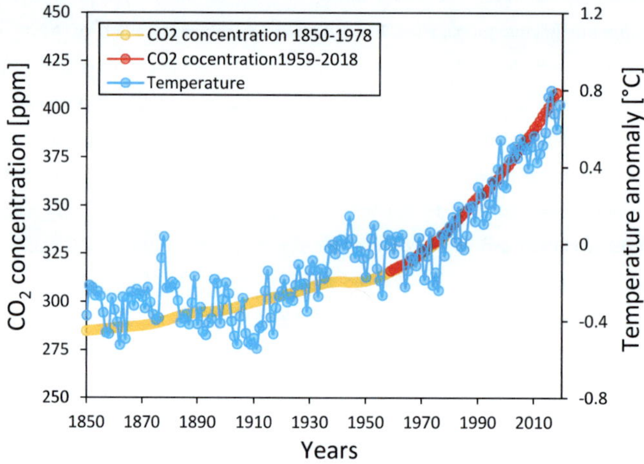

Figure 2.1. Correlation of global temperature and atmosphere CO2 concentration in 1850-2018, Temperature data is from HADCRUTv3, CO2 data is from NOAA ESRL (1859-2018)(NOAA ESRL 2019) (Etheridge et al. 1998) and Low Dome ice core (1850-1978) (Etheridge et al. 1998)

There are many resource of CO$_2$. It can be released through natural and human activities, for instance, volcanic activity, some organic compounds, breathing processes of humans and other animals etc. For a long time, the CO$_2$ concentration in atmosphere is dynamic equilibrium due to plant photosynthesis. Since the beginning of industrial revolution, a large amount of CO$_2$ has been released into atmosphere in a short period, because of massive combustion of fossil fuel, like oil, natural gas and coal etc. Such a huge amount of CO$_2$ breaks the dynamic equilibrium and raises the concentration level of CO$_2$ in atmosphere. This is the main reason for global warming from pre-industrial. How to reduce CO$_2$ releasing and ease the ensuing problems is very popular topic nowadays. Various methods for CO$_2$ storage and utilization has been proposed and projected since 21st century, for example, carbon capture utilization and storage (CCUS), CO$_2$-EOR/EGR, CO$_2$-EGS, CO$_2$ fracturing etc.

2.2. Properties of CO$_2$

As the phase diagram shown in Fig. 2.2, CO$_2$ can exist in the form of solid, liquid and gas phase, respectively, which is associated with temperature and pressure condition. At extreme low temperature and high pressure, it exists as solid phase. With gradually increased temperature, the solid CO$_2$ will transfer to liquid CO$_2$. Exactly on the boundary of solid and liquid phase (melting line), it could exists in a mixture form of solid and liquid phase under equilibrium state. With further decreased pressure, CO$_2$ will transform again from liquid phase to gas phase. On their boundary (saturation line), liquid and gas phase coexist in equilibrium state. Likewise, gas and solid phase could reach the equilibrium state on sublimation line. Especially, there is a point called triple point with temperature of -56.6°C and pressure of 0.52MPa, where is three-phase coexistence (Wu et al. 2005). Besides, there is another special point called critical point (temperature is 31.1°C and pressure is 7.29MPa) (Vargaftik 1975). When the temperature and pressure both exceed them of critical point, CO$_2$ will enter supercritical state, where CO$_2$ owns the characters of both liquid CO$_2$ and gas CO$_2$. Specifically, the supercritical CO$_2$ has a high density close to liquid and low viscosity close to gas, which performs very well as fracturing fluid or working fluid at reducing friction during operation.

The density and viscosity of CO$_2$ is correlated to both pressure and temperature. As shown in Fig. 2.3, the density of CO$_2$ falls down with increased temperature (constant pressure) or decreased pressure (constant temperature). At low temperature (T<10°C) and high pressure (P>350bar), the density of CO$_2$ is comparable with water, even higher. Inversely, at high temperature (T>70°C) and low pressure (P<100bar), the density of CO$_2$ could low as 200kg/m³. The common density of supercritical CO$_2$ from 900kg/m³ to 400kg/m³ shows a high compressibility. In comparison with density, the viscosity shows a comparable correlation with temperature and pressure. In spite of this, the viscosity is very low. The viscosity of supercritical CO$_2$ (0.9×10^{-4} -0.25×10^{-4} Pa·s) is only about 9%-2.5% of water viscosity (about 1.0×10^{-3} Pa·s).

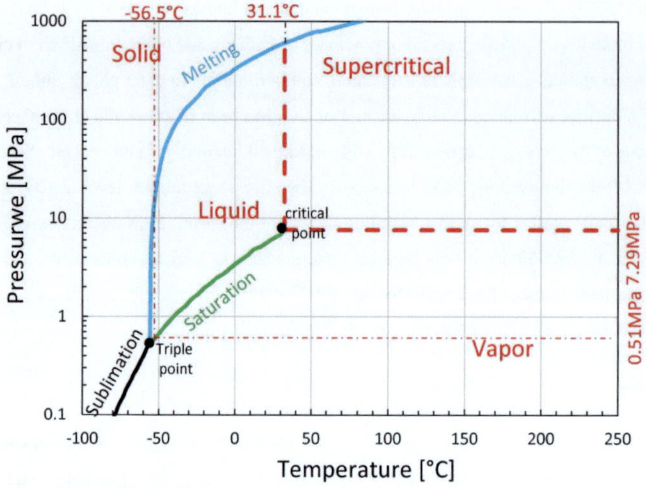

Figure 2.2. Phase diagram of CO₂ (data from Altumin 1975)

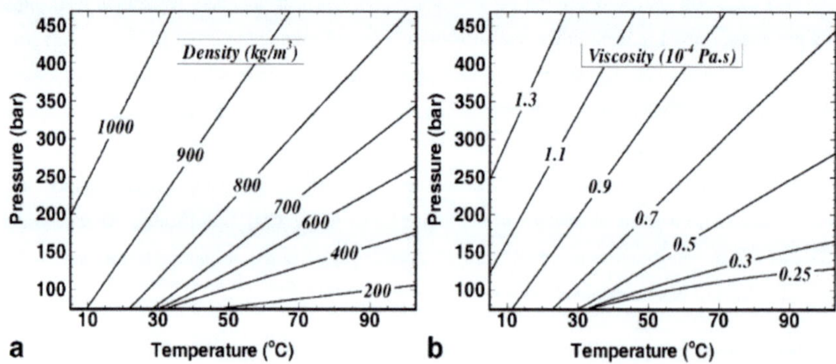

Figure 2.3. Temperature- and pressure-dependent density and viscosity of CO₂ (pressure and Garcia 2002)

The enthalpy of CO_2 is also both temperature- and pressure-dependent. The specific enthalpy of CO_2 and water are compared in Fig. 2.4 under different temperature and pressure. Under the same temperature and pressure, the specific enthalpy per unite mass of water is higher than CO_2. The difference is much significant under high temperature. The specific enthalpy shows a much complex correlation with temperature and pressure, especially around the critical point, where the state of CO$_2$ changes dramatically. This dramatic change also causes that the density as well viscosity varies rapidly even in a narrow temperature and pressure range near the critical point. Such character makes state computation of CO$_2$ very complex.

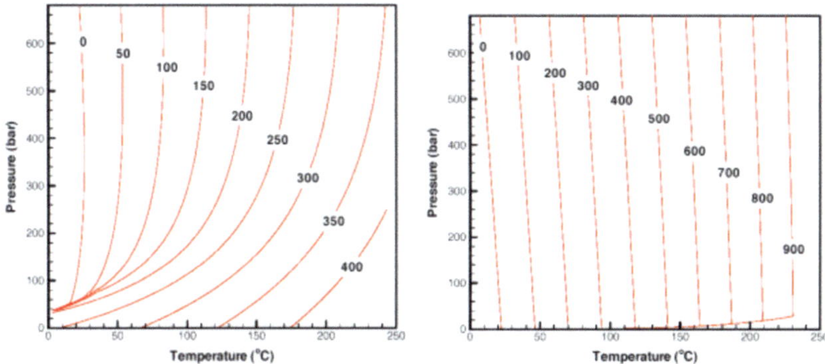

Figure 2.4. Temperature-and pressure-dependent specific enthalpy of CO$_2$ and water in unites of KJ/kg (Pruess 2006)

CO$_2$ is one of acid gas. A series of chemical reactions will take place, after CO$_2$ dissolving in formation water. Firstly, the dissolved CO$_2$ will form carbonic acid H$_2$CO$_3$, which is type of weak acid and its chemical properties is not particularly stable. The carbonic acid H$_2$CO$_3$ will break into HCO$_3^-$ and H$^+$, in which HCO$_3^-$ can further break into CO$_3^{2-}$ and H$^+$. These chemical reactions are given in following, in which the hydrogen ion H$^+$ makes the solution acidic. Besides, all the chemical reactions are reversible. At the beginning, above-mentioned substances reach the chemical dynamic equilibrium. Once other associated substances are incorporated or chemical equilibrium condition is broken, the substance will vary and reaches new equilibrium state. Macroscopically, the process of re-equilibrium performs as precipitation or dissolution mineral, which is the reaction mechanism for long-term CO$_2$ geological storage (Liu 2014).

$$CO_{2(g)} \leftrightarrow CO_{2(aq)} \tag{2.1}$$

$$CO_{2(aq)} + H_2O \leftrightarrow H_2CO_{3(aq)} \tag{2.2}$$

$$H_2CO_{3(aq)} \leftrightarrow HCO_{3(aq)}^- + H_{(aq)}^+ \tag{2.3}$$

$$HCO_{3(aq)}^- \leftrightarrow CO_{3(aq)}^{2-} + H_{(aq)}^+ \tag{2.4}$$

2.3. Phase path of CO_2 in engineering processes

A typical phase paths of CO_2 in a fracturing-treatment cycle in unconventional gas reservoir is illustrated in Fig. 2.5. Position 1 is tank for fracturing. As CO_2 in tank is commonly stored and transported in liquid phase, the Position 1 should locate at least on the saturation line or over the saturation line. Position 2 is pump or wellhead. Before injection, CO_2 need be pressurized by pump. The temperature of CO_2 increases slightly in process 1-2, under the effect of environment temperature. Then the pressurized CO_2 is injected into well bottom or perforation (Position 3). The CO_2 in Position 1, 2, 3 exists mostly in liquid phase, while Position 3 can also exist in supercritical phase, which is affected by many factors, like injection rate, injection depth, formation temperature, heat conductivity etc. Position 4 is hydraulic fracture. In this process 3-4, CO_2 will be heated by heat exchange trough fracture wall, but pressure will fall a bit, for the reason of friction effect of fracture wall roughness. It should be mentioned that the temperature of CO_2 in fracture (Position 4) is not constant, rather increase gradually from perforation to fracture tip. The temperature at fracture tip is infinitely close to formation temperature. Generally, injection well will be shut in for fracture closure and proppant support, in which the temperature increases gradually but still low than formation temperature. After that, CO_2 will reach Position 5, where the CO_2 has a relatively low pressure and high temperature, in comparison with it at Position 4. The CO_2 in both Position 4 and 5 exist normally in supercritical phase. Finally, CO_2 will flow back to surface in gas phase at the ambient condition (Position 6). Especially, in process 3-5, some CO_2 leaks into formation due to pressure difference and exists as supercritical phase in hot reservoir.

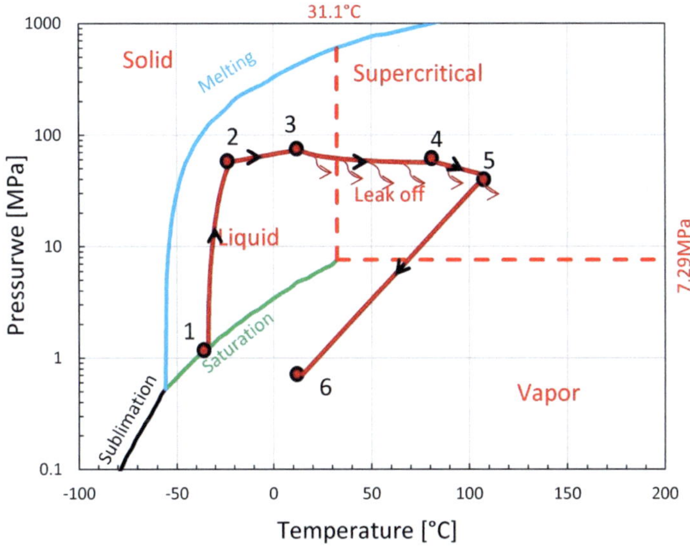

Figure 2.5. Phase diagram of CO$_2$ and typical phase paths CO$_2$ during a fracturing-treatment cycle in unconventional gas reservoir (modified from Li and Zhang 2019)

In exploiting of HDR energy, there are two important engineering processes, including reservoir stimulation and heat extraction. The phase path during reservoir stimulation is comparable with that in unconventional gas reservoir, except for without flow back. The phase paths during heat extraction are illustrated in Fig. 2.6, the process begins at outlet of turbo (Position 1), where the temperature is relative high but pressure is low. Then CO$_2$ will be pressurized at pump or wellhead. In this process, the temperature of CO$_2$ decreases a bit, due to heat loss under ambient temperature. Then the pressurized is piped down to perforation or well bottom at Position 3. Similarly, both temperature and pressure has a small increment comparing them at Position 2, for the influence of well and gravity. In the next step, CO$_2$ flows through HDR reservoir to reach production well, which is represented by Position 4 in Fig. 2.6. In this process, the CO$_2$ heated again by geothermal reservoir will bring the heat out of geothermal reservoir. Similarly, some content of CO$_2$ leaks into formation due to pressure difference. Addition to this, the pressure will decrease under friction effect of fracture roughness. Next, the heated CO$_2$ flows up to the production wellhead (Position 4). Contrary to process (2-3), both pressure and temperature of heated CO$_2$ will decrease in process 4-5. Process 5-1 is a process for energy release, potential process including turbo, room heating, etc., thus the energy release cause a deceased temperature as well as pressure. Other than this, some fresh CO$_2$ need be supplied at Position 1 to balance the leaked content.

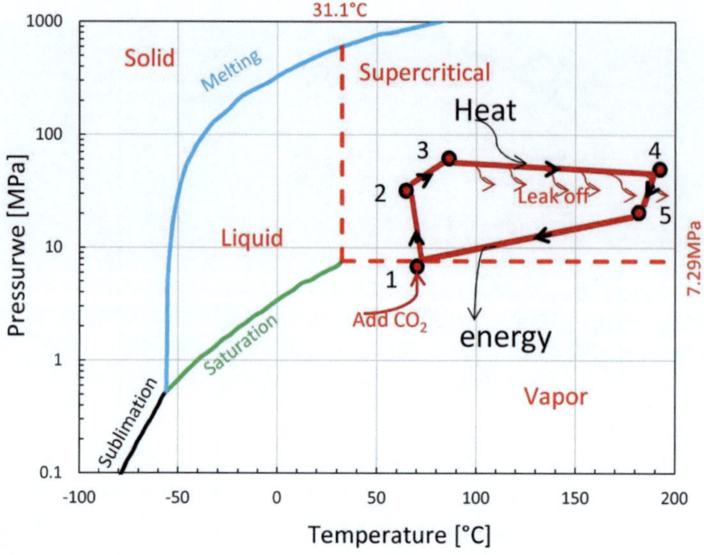

Figure 2.6. Phase diagram of CO$_2$ and typical phase path CO$_2$ during a fracturing-treatment cycle in EGS

2.4. CO$_2$-EGS

2.4.1. Status of CO$_2$-EGS

From the first successful application of EGS in 1997 at the Fenton Hill, USA (Nuckols et al. 1981), it has attracted appreciable attention. Many countries have implemented and planned their own experimental even commercial EGS projects, e.g. Germany (BGR 2019), France (Heimlich et al. 2016; Alber and Solibida 2017), USA (Dobson et al. 2017), Japan (Asanuma et al. 2015), Australia (Mills and Humphreys 2013), China (Feng et al. 2014), Turkey (Liao et al. 2020b), etc. Water as the heat exchange medium has been used in all the EGS projects. Based on the traditional EGS, Brown (2000) proposed a novel CO$_2$-EGS concept, in which the traditional heat exchange medium is replaced by CO$_2$. As its excellent performance of high expansivity and lower viscosity (Pruess 2006, Randolph and Saar 2011), CO$_2$-EGS has received massive attentions around the world in last two decades. Till now, many numerical studies have been carried out in following aspects: (1) CO$_2$ behavior in injection well and production well (Luo et al. 2014, Xu et al. 2015, Liu et al. 2017); (2) performance of CO$_2$ in heat extraction (Pruess 2006, Atrens et al. 2008, Pan et al. 2016). Because of the compressibility/expandability of CO$_2$, a sizeable temperature increase and a temperature decline are observed in injection and production, respectively, which is more significant than almost incompressible water. Besides, the performance of CO_2 in extraction is comparable even better as water. In limited laboratory experiments about CO_2-EGS, the CO_2-water-rock/mineral reaction has been studied mostly (Ueda et al. 2005, Xu and Pruess 2010, Smith et al. 2013, Zhou et al. 2016). The results of Muller et al. (2009) indicated that chemical reaction product of acidic CO_2 dissolution with mineral could block the original flow channel and reduce the conductivity of geothermal reservoir. Maglicoo et al. (2011) performed an experimental study on heat extraction from porous medium by using CO_2. They found that the follow status is flow rate-dependent, diffusion-dominated transients at low flow rates and advection dominated at high flow rates. However, to the best of my knowledge, until now, seldom demonstration projects or large-scale application has been planned around the world.

2.4.2. Competitive advantage

The advantages of CO_2 as working fluid in EGS are discussed primarily in comparison with water-based EGS. Pressure and Azaroual (2006) carried a numerical study to compare the efficiency of CO_2 and water in heat extraction. As shown in Fig. 2.7, the numerical simulation was conducted in a horizontal fracture with height of 50m. Distance between injection and production well is 650m. Some initial parameters, like injection temperature, reservoir temperature and pressure, porosity are shown together in Fig. 2.7. More detailed information can be found in reference (Pruss and Azaroual 2006). The injection and production pressure were fixed at 510bar and 490bar respectively. The results of CO_2 and H_2O as working fluid are plotted in the right of Fig. 2.7. Clearly, CO_2 performs better in heat extraction under same driven pressure (difference between injection and production pressure), because the low viscosity of CO_2 could promote more amount of CO_2 flow through fracture to bring more heat out.

More advantages as well as disadvantages are summarized in Tab. 2.1 in comparison with water. Advantage properties are colored in red, otherwise colored in blue. In produced CO_2, there is no mineral dissolution/precipitation problem, which is the critical risk with potential to block production well in water based EGS. What's more, in the aspects of compressibility/expandability, CO_2 performs better than water. However, the low specific enthalpy will reduce produced energy per unit mass of CO_2. Moreover, the high leakage caused by low viscosity brings another beneficial advantage for CO_2 geological storage.

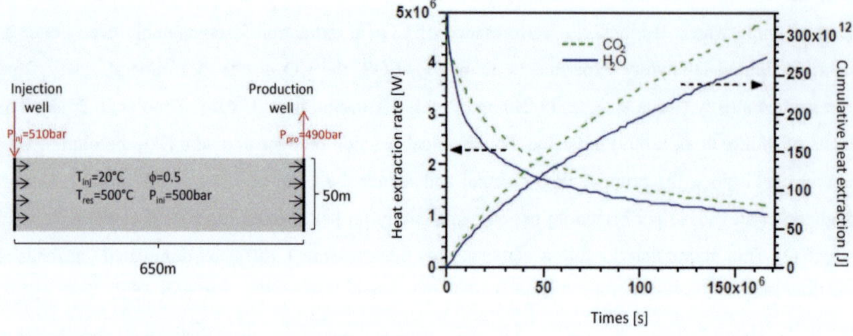

Figure 2.7. Sketch of numerical model, rate of heat extraction and cumulative heat produced for the fracture injection-production problem (modified form Pruess and Azaroual 2006)

Table 2.1. The advantages and disadvantages of CO$_2$-EGS in comparison with water-EGS (adapted from Olasolo et al. 2016)

Fluid properties	CO$_2$	Water
Chemical	Not an ionic dissolution product. No mineral dissolution/precipitation problem	An ionic dissolution product. Serious problem of mineral dissolution/precipitation.
Fluid circulation in wells	High compressible and expandable	Low compressibility. Moderate expandability
Ease of flow in the geothermal reservoir	Low viscosity and density	High viscosity and density
Heat transmission	Low specific heat level	High specific heat level
Fluid loss	Could result in beneficial geological storage of CO$_2$	A hindrance for the development of geothermal reservoir. Costly.

**Advantage properties are colored in red, otherwise colored in blue

2.4.3. Chemical react with mineral and geological sequestration of CO$_2$

In geothermal reservoir, chemical reaction will occur, when the CO$_2$ dissolution encounters mineral under specific condition. The mechanism of such chemical reaction has been studied through injecting CO$_2$ dissolution crossing cores under different condition by Xu et al. (2017). As shown in Fig. 2.8, the chemical reaction occurs dominantly between secondary products (H$^+$, HCO$_3^-$, CO$_3^{2-}$) of CO$_2$

dissolution and other dissolved cation (Ca^{2+}, Mg^{2+}). Therefore, existing of formation water is prerequisite condition for chemical reaction in geothermal reservoir. In addition to this, pressure condition plays also a very important role on chemical reaction. Under increasing pressure, CO$_2$ dissolved process will be accelerated, which further promote mineral dissolution. In this case, the porosity will be enlarged, resulting in lower capillary pressure and high relative permeability. In contrast, a decreased pressure will promote the inverse process, in which more mineral (CaCO$_3$, MgCO$_3$) is precipitated and some CO$_2$ is exsoluted out of CO$_2$ dissolution. The precipitated mineral is potential to block the pores of geothermal reservoir, which shows low relative permeability. pH value is another critical factor effected the chemical reaction in geothermal reservoir. As the pure CO$_2$ dissolution exhibits acidic, any H$^+$ mixing in dissolution will breaks chemical equilibrium. High pH value increases amount of precipitated mineral, while low pH value makes the mineral dissolve in CO$_2$ dissolution.

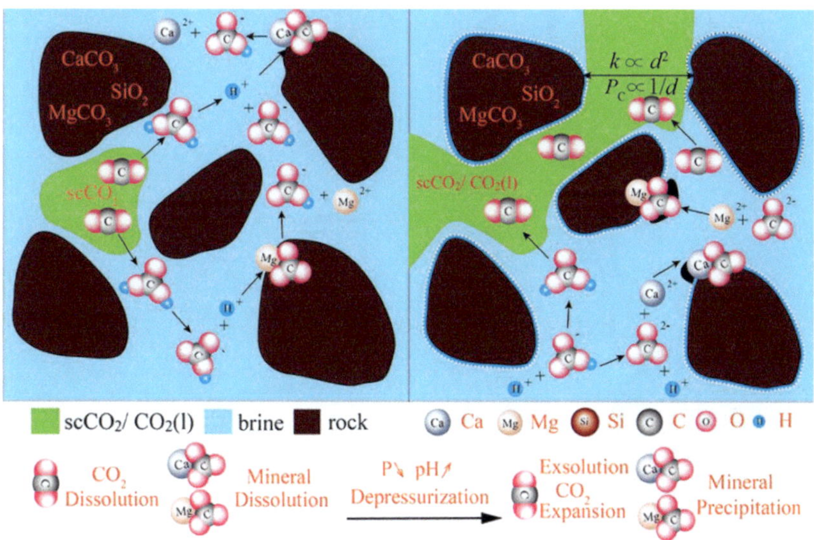

Figure 2.8. Demonstration of mineral dissolution and precipitation (Xu et al. 2017)

In an operated CO$_2$-EGS, the geothermal reservoir can be divided into three main zones according to water saturation, involving zone 1, 2 and 3. In zone 1, there is only supercritical CO$_2$, as the virgin water is replaced by CO$_2$ due to production. Zone 2 distributed surrounding zone 1, in which CO$_2$ exists in gas phase together with water. Outside of zone 2, there is only dissolved CO$_2$ and water, where is dominant region for chemical reaction. Xu and Pruess (2010) conducted numerical study on chemical reaction based on a simple model in zone 3 over 100 years. As shown in Fig. 2.9, the trapped CO$_2$ due to precipitation is illustrated along testing model in different times. The trapped CO$_2$ drops down away from zone 2, which proves that the trapped CO$_2$ is positively correlated to dissolved CO$_2$ concentration. On the other hand, trapped CO$_2$ is accumulated during production and the accumulated rate is slow down production time. The trapped CO$_2$ is treated as long-term CO$_2$ geological storage. The consumption of CO$_2$, namely geological storage, is estimated ranging from 1080kg/MWh to 3960kg/MWh (Brown 2000, Pruess 2006). Frank et al. (2012) conducted a life cycle analysis of GHG emission of combined system (CO$_2$-EGS + traditional coal power) based on the average value of 2520kg/MWh. It is shown that the life cycle emission of combined system was only 15% when 90% carbon capture efficiency was assumed in comparison with traditional coal power.

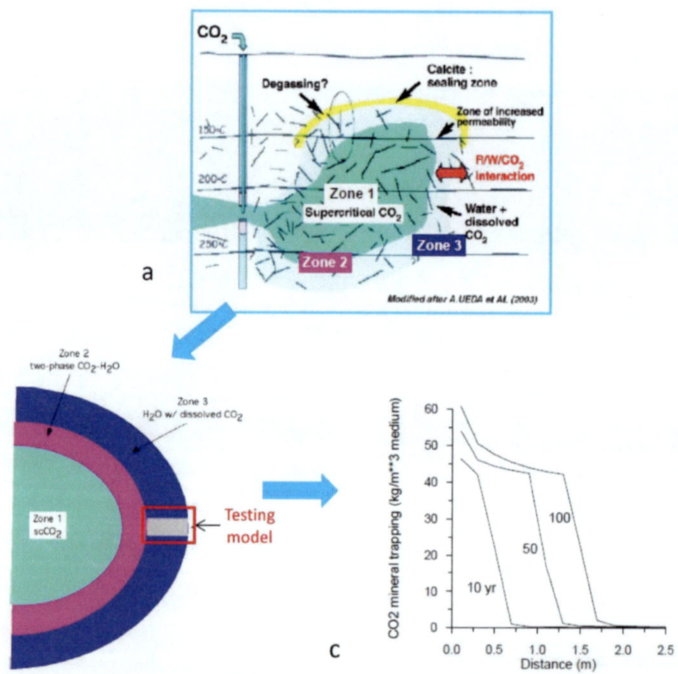

Figure 2.9. (a) Different zones created CO$_2$ injection, (b) sketch of different zones and testing model, (c) cumulative amounts of CO$_2$ sequestered by secondary carbonate precipitation along testing model at different times. (modified from Xu and Pruess 2010)

2.4.4. Challenges of CO_2-EGS

The challenges faced by traditional water based EGS is still affecting the CO_2-EGS, like reservoir stimulation and drilling technology. Because of low permeability, geothermal reservoir, especially HDR reservoir need be enhanced normally by fracturing for efficient heat extraction rate. How to control and evaluate the stimulated results, like fracture morphology, connection with production well, is a challenge. Seismic maps recorded during hydraulic fracturing could be a helpful tool for evaluate fracture morphology (Jeanne et al. 2015, Maurer et al. 2015). Drilling in geothermal reservoir under extremely high temperature and hard rock improves the requirements of drilling technology, maybe resulting into a high investment. Induced seismic is another issue, which observed in many traditional water based EGS (Kraft et al. 2009, Grigoli et al. 2017). CO_2-EGS has also its own challenges. For instance, the pipe as well as surface equipment need be treated against corrosion of acidic CO_2 dissolution. Due to the low viscosity, it is high demands for fracturing equipment and fracturing technology, especially, the tightness of fracturing equipment. The circulation of CO_2 for electricity generation also need be adapted according the properties of CO_2. Then, economic is another noticeable issue that hinders the application of CO_2 EGS.

2.5. Supercritical CO_2 fracturing in unconventional gas reservoir

2.5.1. Status of CO_2 fracturing in unconventional gas reservoir

Over three decades ago, the potential for CO_2 applied as fracturing fluid has been discussed by Lillies (1982) on its properties, like eliminating formation damage caused by water and rapid clean up. Fairless et al. (1986) compared the CO_2 foam fracturing fluid of gelled methanol with other stimulation methods, found that this CO_2 foam performs well for hydrocarbon formation that exhibit low permeability, low reservoir pressure or water sensitivity. Since the successful utilization of CO_2 in American shale formation in early 1990s (Yost et al. 1993), CO_2/sand fracturing treatments have attracted massive attentions of scholars in oil and gas industry. Many research works have been conducted numerically (Yin et al. 2011, Wang et al. 2018, Zhang et al. 2019), experimentally (Ishida et al. 2012, Liu et al. 2018) as well as in field (Gunter et al. 2005, Zhang et al. 2018), to study the behavior of CO_2 as working fluid in unconventional gas reservoir. Comparing with traditional water-based working fluid, CO_2 as fracturing fluid has some excellent features, which is summarized in Tab. 2.2.

Table 2.2. The advantages and disadvantages of CO_2 treatment in unconventional gas reservoir comparing with water-based treatment (adapted from Gandossi (2013) and Middleton et al. (2015))

Fluid properties	CO_2 treatment	Water-based treatment
Formation damage	low	High formation damage in water sensitive formation
Performance in production	Low break pressure, enhance the production due to replacing the adsorbed CH4 by CO_2	High break pressure
Environmental impact	Almost no impact, even migrating CO_2 emission	Pollution of groundwater, flow back water
Water demand	Almost no water demand	About2-4 million gallons per one hydraulic treatment
Flow back	Rapid flow back, minimize flow blocking	Slow, mixing with some substance of hydrocarbon, need be treated or re injection into formation
Investment	High cost	Low cost
Fracture making ability	complex fracture propagation	Planner fracture
Proppant transport	Low	High, especially for gelled fluid
Other	Corrosive due acid solution of CO_2, High leak off	Low corrosive and Low leak off

**Advantage properties are colored in red, otherwise colored in blue

Liquid CO_2 as well as supercritical CO_2 fracturing has a lower break pressure. Meanwhile, due to replacing the adsorbed CH4 in pore by CO_2, the production accumulation of CH_4 can be enhanced in comparison with water-based fracturing fluid. Besides, CO_2 fracturing fluid has a potential to migrate greenhouse emission, instead to pollute groundwater like the chemical additives in water-based fracturing fluid. What's more, it can ease water shortage and minimize the risk for blocking during flow back. Another notable feature is that the CO_2 fracturing fluid can create a complex fracture-network, which is favor for gas production. There are two probable reasons for that: one is the Joule-Thomson effect triggered more cold fractures; another one is the rapid decreased effective stress reactivates the preexisting micro fractures (see in Fig. 2.10). However, CO_2 as fracturing fluid has also some disadvantages, like at present high cost, low proppant-carrying capacity for low viscosity, and corrosive.

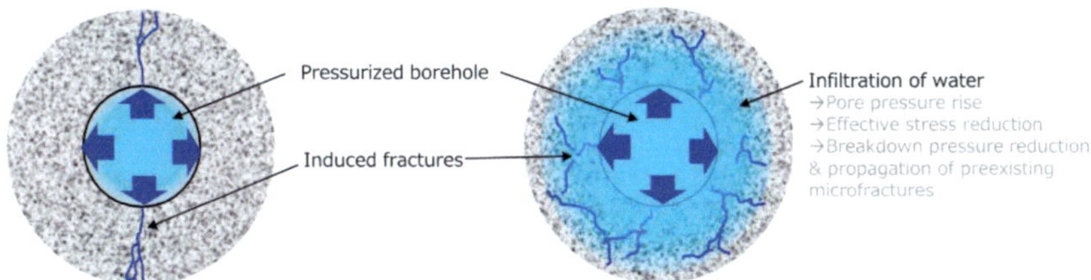

Figure 2.10. Fracture mechanisms for normal fluid and supercritical fluid suggested by Watanabe et al. (2017)

2.5.2. Different CO$_2$ fracturing fluids

According to its composition and state, the CO$_2$ fracturing fluids can be classified into CO$_2$ foam, liquid CO$_2$ and supercritical CO$_2$ fracturing fluid. Their property, low viscosity not only creates more complex fractures, but also causes a significant leakage, especially in high permeable reservoir. Pure CO$_2$ owns a very low viscosity, about several tens to hundreds times low, which limits the permeability range of the reservoir that suitable for pure CO$_2$ fracturing. Generally, in unconventional gas reservoir, not pure CO$_2$, rather the CO$_2$ with improved viscosity by some additives is widely used as fracturing fluid. Various thickened CO$_2$ are recommended by many researchers. The thickener can be classified into three basic categories and some of them are summarized in Tab. 2.3.

Table 2.3. The literature review of different thickened CO_2 in last two decades

Thickener type	Literature	Characteristic
Polymeric thickeners	Bayraktar and Kiran (2000)	That the PDMS was more CO_2-philic than hydrocarbon-based polymers was further confirmed by PDMS solubility data
	Hong et al. (2005)	The polymers were designed and selected to contain one or more sites for favorable thermodynamic inter actions with CO_2
	Wang et al. (2009)	The oligo (3-acetoxy oxetane) (OAO) was less CO_2-philic than PVAc.
	Zhang et al. (2011)	The vinyl ethylether (PVEE) and 1-decene (P-1-D)-thickener could improve the viscosity of CO_2 approximately 13-14 times
	Luo et al. (2015)	The rheological properties and friction performance has been studied with the pressure range from 10-30MPa and temperature range 0-100°C:
	Al Hinai et al. (2018)	The temperature influence on thickened CO_2 has been studied at a high temperature
Small molecule thickeners	Shi et al. (2001)	They synsezized tri (2-perfluorobutylethyl) tin fluoride, (F(CF2)4(CH2)2)3SnF.
	Tapriyal (2009)	They design a non-fluorous thickener for enhancing CO_2 viscosity
	Doherty et al. (2016)	They prepared a series of cyclic amide and urea material as small molecule thickeners for organic solvents, dense CO_2, and the mixtures thereof
	Lee (2017)	A novel small molecules that self-assemble into viscosity enhancing supramolecular structures was proposed for improved mobility control of CO_2 in EOR
Novel thickeners	Paik et al. (2007)	They design a non-fluorous an alog of fluorous bis-ureas.
	Trickett et al. (2010)	They designed surfactants that were not only CO_2-soluble, but also capable of forming viscosity-enhancing rod-like micelles.
	Wu et al. (2018)	They synthesized a CO_2-responsive surfactant,erucamidoproyl dimethylamie (EA) to thicken the CO_2

2.5.3. Challenges of CO_2 fracturing in unconventional gas reservoir

Although many excellent features of CO_2 as fracturing fluid has been pointed out, to date, the CO_2 as fracturing fluid has been studied mostly in lab, only a few field applications have been conducted. Before widely applications, there remain some barriers need be removed. Due to ultra low viscosity, pure CO_2 probably is not efficient to create a fracture for production. Its viscosity could potentially improved by CO_2 thickener. Further, chemical thickener may bring some environmental problems, like traditional fracturing fluid. Proppant transport and placement is another critical factor determining the fracture performance, while CO_2 has a poor proppant-carrying capacity. There are two possible methods to address this problem. Firstly, pure CO_2 is used to create fracture, following proppant transported by traditional fluid. Another method uses thickened CO_2 for fracturing as well as transports light proppant into fracture. Some possible light proppants are listed in Tab. 2.4. Regarding to the overall economic, it depends primarily on its influence on gas production effectiveness as well as additional costs associated with environmental impacts, the economics of CO_2 delivery, and flow-back CO_2 treatment cost (Middleton et al. 2015). As the CO_2 is transported to drilling site commonly in a tank under low temperature and high pressure, it occupies major cost in most cases. Besides, the mixture of CO_2 in produced natural gas is expected in initiate production phase. To meet anti-corrosion requirements for pipeline and engineering equipment, the mixed CO_2 need be removed, which brings additional expense. Let's consider the numerical modeling, because fracturing by using CO_2 more likely to create a complex fracture network, it is a challenge to capture micro-fracture propagation in filed scale model.

Table 2.4. Properties of light proppant introduced by Gaurav et al. (2012)

Properties	ULW-1	ULW-2	ULW-3
Nominal density	1.08	1.25	1.75
Bulk density (g/cc) (closure stress =0 psi)	0.6	0.77	1.19
Porosity of proppant pack (%)	44	36	31
Sphericity	1	0.62±0.7	0.78±0.1

2.6. Enhanced oil and gas recovery using CO_2

2.6.1. Status of CO_2 applications in enhanced oil and gas recovery

The oil production is normally classified into three main stages: primary, secondary and tertiary. In primary stages, the oil production is driven mainly by nature pressure difference between reservoir and production well. Generally, the production rate will slow down with a declined pressure difference. To maintain production rate, a pump system is introduced to keep the pressure difference, which is assigned in secondary stage. However, as shown in Fig. 2.11, the cumulative oil production from primary and secondary recovery is accounted typically only 33% of the total amount of oil present in reservoir. Most of oil is still trapped in reservoir not only for a declined pressure difference, but low mobility due to high viscosity. To address this issue and enhance recovery, two methods, flooding with high temperature steam and CO_2, are widely used. The first field experiment of CO_2 enhancing oil recovery (CO_2-EOR) dates back to 1964 at the Mead Strawn Field, Texas. In this experiment, the efficiency of CO$_2$-EOR has been proven by the result that over 50% more oil was produced using CO$_2$ than by water flooding (Hill et al. 2013). About 10 years late, the first commercial scale was successfully implemented at the SACROC in west Texas (Hill et al. 2013). After that, the commercial application of CO$_2$-EOR increase steady. Currently, CO$_2$-EOR occupies the second place in enhancing oil recovery, only after the thermal flooding. In comparison with EOR, the CO$_2$ enhancing gas recovery (CO$_2$-EGR) is a relatively new concept (Hou et al. 2011, Hou et al. 2012), which is mostly associated with CO$_2$ geological sequestration. Because natural gas has an extremely low viscosity in comparison with oil in reservoir, the trapped natural gas in depleted gas reservoir is too small amount to attract attention. But if it treated as secondary production of CO$_2$ sequestration, the produced CO$_2$ could further reduce the cost of CO$_2$ sequestration.

In CO$_2$-EOR, the oil mixed with CO$_2$ improves the mobility significantly. This is why injection of CO$_2$ into oil reservoir could enhance oil recovery (Holm 1986). According to Perera et al. (2016), there are two mixing behaviors of CO$_2$ and oil, miscibility and solubility, which is dependent on reservoir pressure and temperature. Miscible oil recovery is applied only at pressure higher than the minimum miscibility pressure, usually in deep oil reservoir, while immiscible oil recovery is applied to moderately heavy oil reservoir and shallower light oil reservoir. In CO$_2$-EGR, the CO$_2$ injection shows a different mechanism to enhance gas recovery. The absorbed CH$_4$ can be replaced and flooded to production well, through competitive adsorption of injecting CO$_2$.

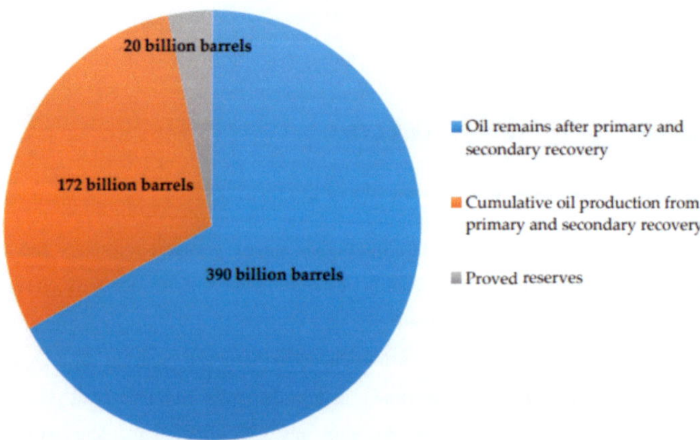

Figure 2.11. Oil production ability by primary/secondary recovery (Kuuskraa et al. 2006)

2.6.2. Challenges of CO$_2$ enhanced oil and gas recovery

Though the CO$_2$-EOR has a long history and is widely used, there remain some problems for commercial application. Because of the mobility difference between CO$_2$ and oil, it is more likely for the injecting CO$_2$ to cross the oil flied forming a finger shape, which reduces the efficiency of EOR significantly. There are two potential solutions: firstly, using a water-alternating-gas injection strategy reduces the risk of forming finger channel potentially, as shown in Fig. 2.12; and CO$_2$ thickener is introduced to increase the viscosity and flooding performance of CO$_2$ in EOR (Enick et al. 2012). The cost is the key factor for CO$_2$-EOR application, which is strongly dependent on the prevailing oil price as well as availability of cheap source of CO$_2$. In addition to this, the unexpected geologic heterogeneity in CO$_2$-EOR brings more potential risks. Regarding to CO$_2$-EGR, the economics is also an obstacle for it application. Besides, the expensive and recovery reduction due to gas mixing is another unsolved problem (Oldenburg 2003). But if consider the benefit gathered by CO$_2$ sequestration, it has potential to be widely used in the further.

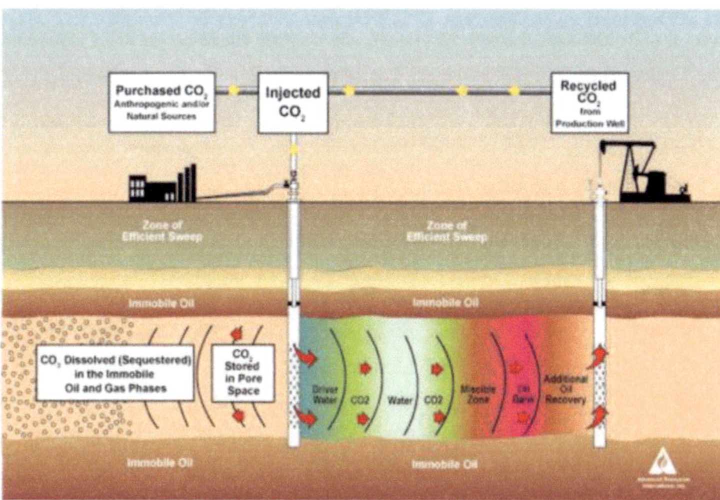

Figure 2.12. Demonstration of water-alternating-gas (Godec et al. 2013)

3. Fundamentals of the THM model

Engineered application of CO_2 in geo-energy production is a very complicated process, containing: multi-component and flow, like CO_2, aqueous and heat flow in EGS, and CO_2, CH_4, water, guar gum, water in unconventional gas reservoir during fracturing; proppant transport and placement; fracture propagation; altering reservoir permeability and porosity, etc. All these processes need be generated on the fundament of thermal-hydraulic-mechanic (THM) coupled framework, which will be introduced in detail in this chapter. The THM coupled framework was developed on the base of coupling concept FLAC3D-TOUGH (Rutqvist and Tsang 2002, Rutqvist et al. 2002). Afterwards, to improve the computing efficient, the TOUGH has been extended to multiple processing programs TOUGH2M and the coupling has been integrated on platform of WinStock (Gou 2018). TMVOC EOS (Pruess and Battistelli 2002) is adopted in this THM coupled framework to extend the scope of temperature of mixtures. Addition to this, more available components and phases provide the opportunity for user to define new component. For an instance, guar gum concerned as a hydrate has been defined in this work according its rheological characteristics. In this chapter, the fundamental of multi-component and flows, basic mechanics as well as the THM coupled mechanism are introduced in following.

3.1. Hydraulic flow in porous medium

3.1.1. Multi-phase and -component flow in porous medium

Maximum four phases are involved in this work, including gaseous, aqueous, hydrate and solid phases, in which the first three phase is mobile phase, comprising of CO_2, CH_4, water and guar gum respectively. As shown in Fig. 3.1, the mobile phases exist in the void space of this solid matrix. In a representative zone, local thermal equilibrium condition is considered, namely the solid and the neighboring fluid are at the same temperature, as well the ice formation is neglected. The flows of all mobile phase obey Darcy Law in pore medium. Gaseous, aqueous and hydrate phase are treated as ideal mixture, which is distributed uniformly in the void of a representative zone. The proportion of each phase is defined as average mole fraction in the phase over all phases in the representative zone. Therefore, the sum of all mobile phase is equal to 1 (Eq. 3.1).

$$\sum_\beta S_\beta = 1 \tag{3.1}$$

Where S_β is the mole fraction in phase β [-]; β is the mobile phase, including gas phase (g), aqueous phase (w) and hydrate phase (h). In these phases, the proportion of each component in every phase is also represented by the average mole fraction of component over this phase. Similarly, their sum is 1 (Eq. 3.2)

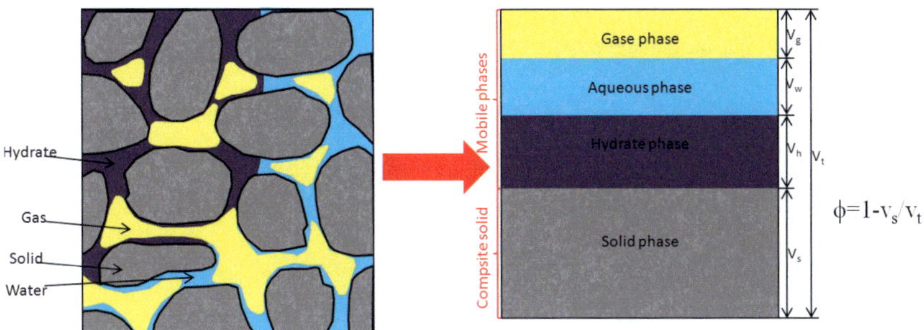

Figure 3.1. Demonstration of multi-component in a representative zone

$$\sum_\kappa x_\beta^\kappa = 1 \tag{3.2}$$

Here x_β^κ is the mole fraction of component κ in phase β [-]. The transport process, characterizing the multi flow in geothermal reservoir as well as in unconventional gas reservoir, can be described by invoking the conservation law for mass and energy. The mass conservation for each component is given in following equation

$$\frac{\partial M^\kappa}{\partial t} = -\nabla \cdot \boldsymbol{F}^\kappa + q^\kappa \tag{3.3}$$

M^κ is the mass accumulation term [mol/m³]; $\boldsymbol{F}^\kappa$ is the flux term [mol/m²/s]; q^κ is the accumulated source term [mol/m³/s]. They are given in following equations, respectively.

$$M^\kappa = \sum_\beta \phi S_\beta \rho_\beta X_\beta^\kappa \tag{3.4}$$

$$\boldsymbol{F}^\kappa = \sum_\beta X_\beta^\kappa \rho_\beta \boldsymbol{u}_\beta \tag{3.5}$$

$$q^\kappa = \sum_\beta q_\beta^\kappa \tag{3.6}$$

Where ϕ is the porosity [-]; ρ_β is the mole density of mixture in phase β [mol/m³]; q_β^κ is the source term in phase β [mol/m³/s], $\boldsymbol{u}_\beta$ is the volume flux in phase β [m/s]. This volume flux can be expressed by following Darcy law.

$$\boldsymbol{u}_\beta = -\frac{k\, k_{r\beta}}{\mu_\beta}(\nabla P_\beta - \rho_{m\beta}\boldsymbol{g}) \tag{3.7}$$

The k is the permeability [m²], $k_{r\beta}$ is the relative permeability relating to phase β [-]; μ_β is the phase viscosity [Pa·s]; ∇P_β is the pressure gradient in phase β [Pa/m]; $\rho_{m\beta}$ is the mass density of mixture in phase β [kg/m³]; g is the gravity [m/s²].

Substituting Eqs. 3.4-3.6 into Eq.3.3, the mass conservation can be transformed into

$$\Sigma_\beta \left[\partial t (\phi S_\beta \rho_\beta X_\beta^\kappa) \right] + \Sigma_\beta \left[\nabla \cdot (X_\beta^\kappa \rho_\beta \boldsymbol{u}_\beta) \right] - \Sigma_\beta q_\beta^\kappa = 0 \tag{3.8}$$

Combing with Eq. 3.7, the Eq. 3.8 is derived into

$$\Sigma_\beta \left[\partial t (\phi S_\beta \rho_\beta X_\beta^\kappa) \right] + \Sigma_\beta \nabla \cdot \left\{ X_\beta^\kappa \rho_\beta \left(-\frac{k\, k_{r\beta}}{\mu_\beta} (\nabla P_\beta - \rho_\beta \boldsymbol{g}) \right) \right\} - \Sigma_\beta q_\beta^\kappa = 0 \tag{3.9}$$

Then Eq. 3.9 can be expanded further into following equation based on different phases.

$$\partial t \left[\phi \left(S_g \rho_g X_g^\kappa + S_l \rho_l X_l^\kappa + S_h \rho_h X_h^\kappa \right) \right] + \nabla \cdot \left\{ X_g^\kappa \rho_g \left[-\frac{k\, k_{rg}}{\mu_g} (\nabla P_g - \rho_{mg} \boldsymbol{g}) \right] + X_l^\kappa \rho_l \left[-\frac{k\, k_{rl}}{\mu_l} (\nabla P_l - \right. \right.$$
$$\left. \left. \rho_{ml} \boldsymbol{g}) \right] + X_h^\kappa \rho_h \left[-\frac{k\, k_{rh}}{\mu_h} (\nabla P_h - \rho_{mh} \boldsymbol{g}) \right] \right\} - \left(q_g^\kappa + q_l^\kappa + q_h^\kappa \right) = 0 \tag{3.10}$$

Where ρ_{mg}, ρ_{ml} and ρ_{mh} are the mass densities in gas, aqueous and hydrate phase [kg/m³].

3.1.2. Heat conservation

Under the assumption of local thermal equilibrium, the energy conservation in porous medium can be described by one energy balance equation (Eq. 3.11), which has a similar formula as Eq. 3.3.

$$\frac{\partial M}{\partial t} = -\nabla \cdot \boldsymbol{F} + q \tag{3.11}$$

M is heat accumulation term [J/m³]; $\boldsymbol{F}$ is heat flux [J/m²/s]; q is heat resource [J/m³/s].

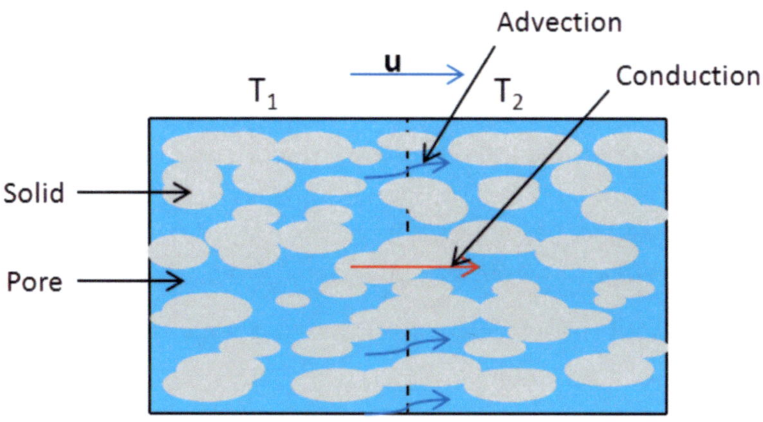

Figure 3.2. Illustration of heat advection and conduction

The three terms are given in following equations, in which the heat accumulation term is composed of two parts; the heat accumulation of solid and multi fluid in pore of representative zone. In Addition to this, heat flux is consisted of two different terms as well; one is advection terms due to multi-flows through pores, another one is conduction term due to heat flow through solid under a specific temperature difference. As shown in Fig. 3.2, the heat advection owns a direction same as flow direction in pores, while heat conduction is driven by temperature gradient.

$$M = (1 - \phi)\rho_s U_s + \sum_\beta \phi S_\beta \rho_\beta U_\beta \tag{3.12}$$

$$F = \sum_\beta \rho_\beta u_\beta h_\beta - k^c_{eff} \nabla T \tag{3.13}$$

$$q = \sum_\beta q_\beta h_\beta \tag{3.14}$$

Where ρ_s is the mass density of solid [kg/m³]; h_β is the mole enthalpy [J/mol]; k^c_{eff} is effective thermal conductivity [J/°C/mol]; U_s is the specific intern energy of solid [J/kg]; U_β is specific intern energy of mobile phase g, w and h [J/mol]. The defective thermal conductivity and heat capacity can be expressed by following equations.

$$k^c_{eff} = (1 - \phi)k^c_s + \sum_\beta \sum_\kappa \left(\phi X^\kappa_\beta S_\beta k^c_\beta\right) \tag{3.15}$$

$$U_s = \int_{T_{ref}}^{T} C_{v_s} dT \tag{3.16}$$

$$U_\beta = \int_{T_{ref}}^{T} C_{v_\beta} dT \tag{3.17}$$

C_{v_s} is specific heat capacity at constant volume of solid [J/°C/kg]; C_{v_β} is specific heat capacity at constant volume of mobile phase g, w and h [J/°C/mol]; k_s^c is thermal conductivity of solid [J/m²/s/°C]; k_β^c is thermal conductivity of mobile phase g, w and h [J/m²/s/°C];

Substitute Eqs. 3.7 & 12-14 into Eq. 3.11, the heat conservation can be extended into following

$$\partial t\left[(1-\phi)\rho_s U_s + \phi\left(S_g\rho_g U_g + S_l\rho_l U_l + S_h\rho_h U_h\right)\right]$$
$$+ \nabla\left\{h_g\rho_g\left[-\frac{k\,k_{rg}}{\mu_g}(\nabla P_g - \rho_{mg}\boldsymbol{g})\right] + h_l\rho_l\left[-\frac{k\,k_{rl}}{\mu_l}(\nabla P_l - \rho_{ml}\boldsymbol{g})\right]\right.$$
$$\left. + h_h\rho_h\left[-\frac{k\,k_{rh}}{\mu_h}(\nabla P_h - \rho_{mh}\boldsymbol{g})\right]\right\} - k_{eff}^c\nabla T - \left(q_g h_g + q_l h_l + q_h h_h\right) = 0 \qquad (3.18)$$

3.1.3. Space and time discretization

To solve multi-flow in porous medium, the whole model need be discretized firstly into small blocks using the integral finite difference method (Narasimhan and Witherspoon, 1976). Fig. 3.3 shows a discrete block with a sub-domain V_n of the flow system, which is bounded by the closed surface Γ_n. Therefore, the integral of Eq. 3.3 over a block can be written in following equations.

$$\frac{d}{dt}\int_{V_n} M^\kappa dV_n = -\int_{\Gamma_n} F^\kappa \cdot \boldsymbol{n}\, d\Gamma_n + \int_{V_n} q^\kappa dV_n \qquad (3.19)$$

Where $\boldsymbol{n}$ is the normal unite direction [-].

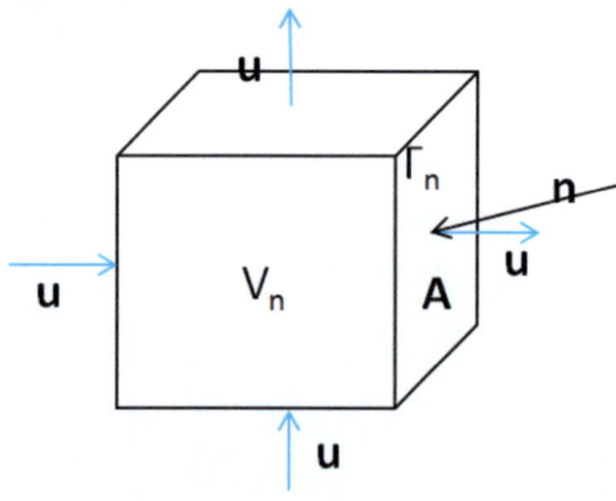

Figure 3.3. Demonstration of a discrete block

It is assumed that different components as well as source distributes uniformly in one control volume of a block. Introducing appropriate volume averages, the integral of Eq. 3.19 can be sampled into following equation.

$$\frac{dM_n^\kappa}{dt} = \frac{1}{V_n}\int_{\Gamma_n} F^\kappa \cdot \boldsymbol{n}\, d\Gamma_n + q^\kappa \tag{3.20}$$

Time is discretized as a first-order finite difference. The flux and source terms in Eq. 3.20 are evaluated at the new time step ($t + 1 = \Delta t + t$) to achieve numerical stability needed for an efficient calculation of multiphase flow. Because the fluxes on the new time step is expressed in terms of the unknown thermodynamic parameters, these unknowns are only implicitly defined in the resulting equations. The content variation can be approximated by

$$M_n^{\kappa,t+1} - M_n^{\kappa,t} = \frac{\Delta t}{V_n}\int_{\Gamma_n} F^{\kappa,t+1} \cdot \boldsymbol{n}\, d\Gamma_n + \Delta t q^{t+1} \tag{3.21}$$

Here an residual function is introduced to describe the content difference at time step $t + 1$.

$$R_{ij}^{\kappa,t+1} = M_{ij}^{\kappa,t+1} - M_{ij}^{\kappa,t} - \left\{\frac{\Delta t}{V_{ij}}\int_{\Gamma_{ij}} F^{\kappa,t+1} \cdot \boldsymbol{n}\, d\Gamma_{ij} + \Delta t q^t\right\} = 0 \tag{3.22}$$

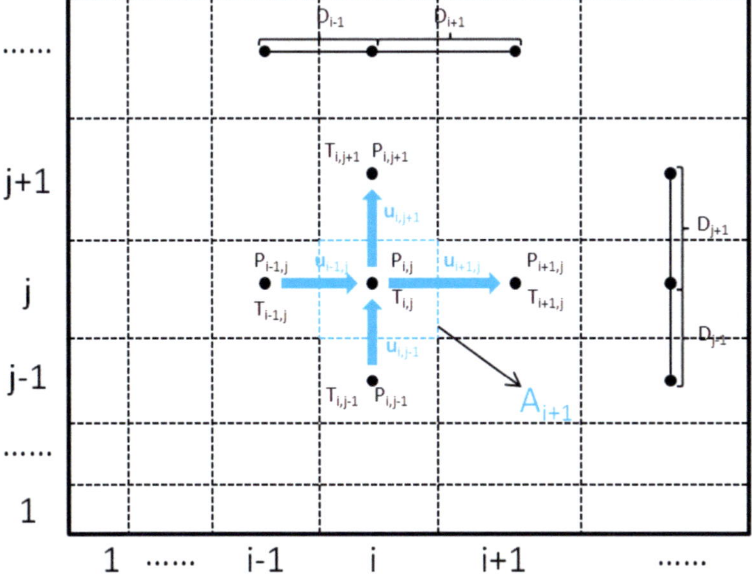

Figure 3.4. Topology of a 2-diemsional discrete space

The integral of flux on closed surface in 2D discrete space (as shown in Fig. 3.4) can be written is following equation.

$$\int_{\Gamma_n} F^{\kappa,t+1} \cdot n \, d\Gamma_n = \sum_\beta X_\beta^{\kappa,t+1} \rho_\beta^{t+1} \left(A_{i-1} u_{\beta,i-1,j}^{t+1} - A_{i+1} u_{\beta,i-1,j}^{t+1} + A_{j-1} u_{\beta,i,j-1}^{t+1} - A_{j+1} u_{\beta,i,j+1}^{t+1} \right) \quad (3.23)$$

Where $A_{i-1}, A_{i+1}, A_{j-1}, A_{j+1}$ are the contact area in left, right bottom and top sides [m²]; $u_{\beta,i-1,j}^{k+1}$, $u_{\beta,i+1,j}^{k+1}$, $u_{\beta,i,j-1}^{k+1}$, $u_{\beta,i,j+1}^{k+1}$ are the flux from the left, right, bottom and top sides at time step $t + 1$ [m/s].

Likewise, the residual function of heat on a block is introduced as following equation.

$$R_{ij}^{t+1} = M_{ij}^{t+1} - M_{ij}^{t} - \left\{ \frac{\Delta t}{V_{ij}} \int_{\Gamma_{ij}} F^{t+1} \cdot n \, d\Gamma_{ij} + \Delta tq \right\} = 0 \quad (3.24)$$

Eq. 3.24 in a 2D discrete space can be written

$$\int_{\Gamma_n} F^{t+1} \cdot n \, d\Gamma_n = \sum_\beta \rho_\beta^{t+1} \left(A_{i-1} u_{\beta,i-1,j}^{t+1} h_{\beta,?}^{t+1} - A_{i+1} u_{\beta,i+1,j}^{t+1} h_{\beta,?}^{t+1} + A_{j-1} u_{\beta,i,j-1}^{t+1} h_{\beta,?}^{t+1} - $$
$$A_{j+1} u_{\beta,i,j+1}^{t+1} h_{\beta,?}^{t+1} \right) + k_{eff}^c \left(T_{i-1,j}^{t+1} + T_{i+1,j}^{t+1} + T_{i,j-1}^{t+1} + T_{i,j+1}^{t+1} - 4 T_{i,j}^{t+1} \right) \quad (3.25)$$

$h_{\beta,?}^{t+1}$ is the mixture enthalpy in phase β [J/mol], which is related to the flux direction of phase β. It can be determined according to Eq. 3.26.

$$h_{\beta,?}^{t+1} = \begin{cases} h_{\beta,i-1,j}^{t+1} & if \ u_{\beta,i-1,j}^{t+1} \geq 0 \\ h_{\beta,i,j}^{t+1} & if \ u_{\beta,i-1,j}^{t+1} < 0 \end{cases} \quad (3.26)$$

The flux of element ij with surrounding elements is given in Eq.3.27, which is dependent on current pressure.

$$\begin{cases} u_{\beta,i-1,j}^{t+1} = \frac{k \, k_{rh}}{\mu_\beta} \left(\frac{P_{\beta,i-1,j}^{t+1} - P_{\beta,i,j}^{t+1}}{D_{i-1}} - \rho_{m\beta}^{t+1} g \cdot n \right) \\ u_{\beta,i+1,j}^{t+1} = \frac{k \, k_{rh}}{\mu_\beta} \left(\frac{P_{\beta,i,j}^{t+1} - P_{\beta,i+1,j}^{t+1}}{D_{i+1}} - \rho_{m\beta}^{t+1} g \cdot n \right) \\ u_{\beta,i,j-1}^{t+1} = \frac{k \, k_{rh}}{\mu_\beta} \left(\frac{P_{\beta,i,j-1}^{t+1} - P_{\beta,i,j}^{t+1}}{D_{j-1}} - \rho_{m\beta}^{t+1} g \cdot n \right) \\ u_{\beta,i-1,j}^{t+1} = \frac{k \, k_{rh}}{\mu_\beta} \left(\frac{P_{\beta,i,j+1}^{t+1} - P_{\beta,i,j}^{t+1}}{D_{j+1}} - \rho_{m\beta}^{t+1} g \cdot n \right) \end{cases} \quad (3.27)$$

NK components are considered in this work. Under isothermal condition, the number of equations (NEQ=NK) is equal to the number of components. Under non-isothermal condition, one more equation of heat flow is taken into account (NEQ=NK+1). In the whole model, the space is divided into NEL individual elements. In each element, there are NEQ residual equations are constructed. Obviously, the residual function for multi-component as well as heat is none linear. Therefore, the Newton/Raphson iteration method is introduced to solve this none linear equation. The general formula is written as following equation.

$$\mathbf{X}^{(p+1)}=\mathbf{X}^{(p)}-\mathbf{J}^{-1}\mathbf{R} \tag{3.28}$$

Where $\mathbf{X}^{(p)}$ is the variable matrix at iteration step (p); $\mathbf{X}^{(p+1)}$ is the variable matrix at iteration step (p+1); $\mathbf{R}$ is the residual matrix; $\mathbf{J}$ is the Jacobi matrix based on the residual matrix.

Both $\mathbf{X}$ and $\mathbf{R}$ matrix is composed of NEQ×NEL variables, whose correlation with the element is illustrated in Fig. 3.5. Jacobi matrix is formed by derivation of residual function to the variable in $\mathbf{X}$ matrix respectively. In consequence, total NEQ×NEL rows and columns are built in Jacobi matrix $\mathbf{J}$. Replacing these matrixes and variables by specific variables, Eq. 3.28 can be written in Eq. 3.29.

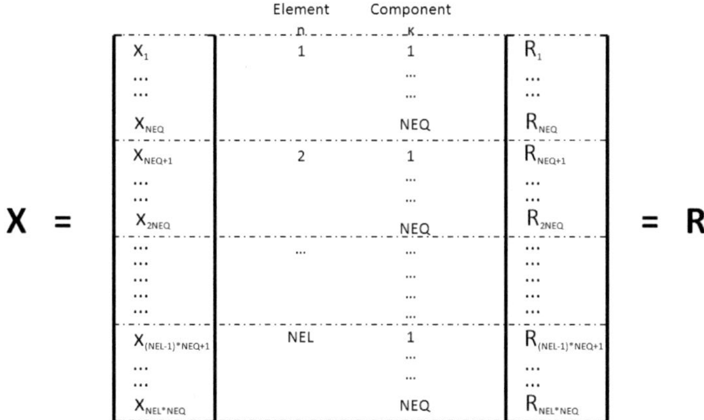

Figure 3.5. Correspondence between variables and residual function to element

$$\begin{bmatrix} x_1^{(p+1)} \\ x_2^{(p+1)} \\ ... \\ x_{NEL\times NEQ}^{(p+1)} \end{bmatrix} = \begin{bmatrix} x_1^{(p)} \\ x_2^{(p)} \\ ... \\ x_{NEL\times NEQ}^{(p)} \end{bmatrix} - \begin{bmatrix} \dfrac{\partial R_1}{\partial x_1} & \dfrac{\partial R_1}{\partial x_2} & ... & \dfrac{\partial R_1}{\partial x_{NEL\times NEQ}} \\ \dfrac{\partial R_2}{\partial x_1} & \dfrac{\partial R_2}{\partial x_2} & ... & \dfrac{\partial R_2}{\partial x_{NEL\times NEQ}} \\ ... & ... & ... & ... \\ \dfrac{\partial R_{NEL\times NEQ}}{\partial x_1} & \dfrac{\partial R_{NEL\times NEQ}}{\partial x_2} & ... & \dfrac{\partial R_{NEL\times NEQ}}{\partial x_{NEL\times NEQ}} \end{bmatrix}^{-1} \begin{bmatrix} R_1 \\ R_2 \\ ... \\ R_{NEL\times NEQ} \end{bmatrix} \tag{3.29}$$

To achieve the variable at new time step (t + 1), the initial variables (initial at beginning t_0 or form last time step t) is necessary for start the iteration process. After n step iterations, the value of residual falls to a very low level. When the residual satisfied the convergence tolerance that defined in Eq. 3.30 or Eq. 3.31, the variables gained in last iteration step are treated as the variables at new time step.

However, if convergence cannot be achieved within a certain number of iterations, the time step Δt will be reduced for a new iteration process.

$$\left|\frac{R_{ij}}{M_{ij}}\right| \leq \varepsilon_1 \tag{3.30}$$

ε_1 is the relative convergence tolerance [-].

If the accumulation terms is too small (low as 1), another other parameters named absolute convergence tolerance are introduced to re-defined the convergence limits in Eq. 3.31.

$$\left|R_{ij}\right| \leq \varepsilon_1 \cdot \varepsilon_2 \tag{3.31}$$

3.1.4. Equations of state

The primary parameters like temperature and pressure are gained by solving above-mentioned multi-flow equations. On bases of these primary parameters, volume or density of fluid still as unknown can be solved according to the relationship between volume, pressure and temperature of the fluid, which is called equation of state (EOS). Many EOSs has been developed to solve the fluid with different composition. In the hypotheses of ideal gas, namely without considering molecule force, the volume, pressure and temperature of gas shows a linear correlation, which is expressed in Eq. 3.32.

$$p = \frac{nRT}{V} \tag{3.32}$$

Where n is the number of moles of the gas [mol]; R is the universal gas constant [w/mol/°C].

The hypotheses of ideal gas limit its application. For real gas, the molecule force plays a very important role, especially under compression with a high pressure. A most fundamental EOS named Van der Waals EOS was proposed by Van der Waals in 1873. The molecule force is considered by introducing two addition parameters, a for attraction and b for repulsion effect of molecule force. This cubic law can be used to predict the gas and liquid phase. Yet, the predicting accuracy sometimes is not very well.

$$p = \frac{RT}{V-b} - \frac{a}{V^2} \tag{3.33}$$

There are two very popular equation of state in petroleum engineering, Soave-Redlich-Kwong EOS (SRK) (Soave 1972) and Peng Robinson EOS (PR) (Peng and Robinson 1976).

SRK EOS:

$$p = \frac{RT}{V-b} - \frac{a}{V(V+b)} \tag{3.34}$$

or

$$Z^3 - Z^2 + (A - B - B^2) \cdot Z - A \cdot B = 0 \tag{3.35}$$

Where Z is compressibility factor [-]; other constants for SRK EOS are defined as following equations.

$$A = a \cdot \frac{p}{(RT)^2} \quad ; \quad B = b \cdot \frac{p}{(RT)} \tag{3.36}$$

$$a = 0.42748 \cdot \frac{(RT_c)^2}{p_c} \cdot \alpha \quad ; \quad b = 0.08664 \cdot \frac{(RT_c)}{p_c} \tag{3.37}$$

$$\alpha = \left[1 + m \left(1 - \sqrt{\frac{T}{T_c}} \right) \right]^2 \tag{3.38}$$

$$m = 0.48 + 1.574\omega - 0.176\omega^2 \tag{3.39}$$

Where ω is acentric factor; T_c is critical temperatur [K], P_c is critical pressure [Pa].

PR EOS:

$$p = \frac{RT}{V-b} - \frac{a}{V(V+b)+b(V-b)} \tag{3.40}$$

or

$$Z^3 - (1 - B)Z^2 + (A - 2B - 3B^2) \cdot Z - (AB - B^2 - B^3) = 0 \tag{3.41}$$

The constants for PR EOS are defined as following equations.

$$A = a \cdot \frac{p}{(RT)^2} \quad ; \quad B = b \cdot \frac{p}{(RT)} \tag{3.42}$$

$$a = 0.45724 \cdot \frac{(RT_c)^2}{p_c} \cdot \alpha \quad ; \quad b = 0.0778 \cdot \frac{(RT_c)}{p_c} \tag{3.43}$$

$$\alpha = \left[1 + m \left(1 - \sqrt{\frac{T}{T_c}} \right) \right]^2 \tag{3.44}$$

$$m = 0.37464 + 1.54226\omega - 0.26992\omega^2 \tag{3.45}$$

As introducing of parameters a and b, both SRK and PR EOS could be used to predict the phase change of mixture. According to Eqs. 3.37 & 3.43, both a and b are measured relating to the critical temperature and critical pressure. Comparing with the typical range 0.24-0.29 of real fluid, the critical compressibility factor of Van der Waal of 0.375 is too high. SRK EOS has a relative low critical compressibility factor (0.333), which makes it perform well in lighter hydrocarbon. However, for heavier hydrocarbon, PR EOS works better than SRK EOS as its critical compressibility factor low as 0.3074. Normally, three roots are obtained by solving the above-mentioned cubic EOS. The smallest root is used to evaluate liquid density according to Eq. 3.46, as liquid fluid has higher density and is difficult compressed in comparison with gases phase. On the contrast, the biggest one is used to evaluate the gas density.

$$\rho = \frac{p}{RTZ} \tag{3.46}$$

A mixture has many components, whose properties are dependent not only on pressure and temperature, but on composition. Therefore, the composition should be considered in the mixing rule for calculating mixing properties. A simple mixing rule based on molar-weighted summation is adopted in this work. The average molar mass of a mixture is given in following

$$\overline{M} = \frac{1}{\sum_i \frac{x_i}{M_i}} = \sum_i y_i M_i \tag{3.47}$$

In the mixture, the gases are assumed to behavior similarly under the same corresponding state. In this case, the critical temperature and pressure of mixture also can be calculated by simple weighted summation (Eq. 3.48).

$$p_{pc} = \sum_i y_i \, p_{ci} \, , \quad T_{pc} = \sum_i y_i \, T_{ci} \tag{3.48}$$

Here p_{pc}, T_{pc} are pseudo critical pressure and temperature of mixture. Then the state of mixture are expressed by pseudo reduced pressure (p_{pr}) and temperature (T_{pr}) as

$$p_{pr} = \frac{p}{p_{pc}} \, , \quad T_{pr} = \frac{T}{T_{pc}} \tag{3.49}$$

The relationship between compressibility factor and real gas state (pseudo reduced pressure and temperature) is exhibited in Fig. 3.6, which is a widely used as guide to determine the volumetric properties of mixture.

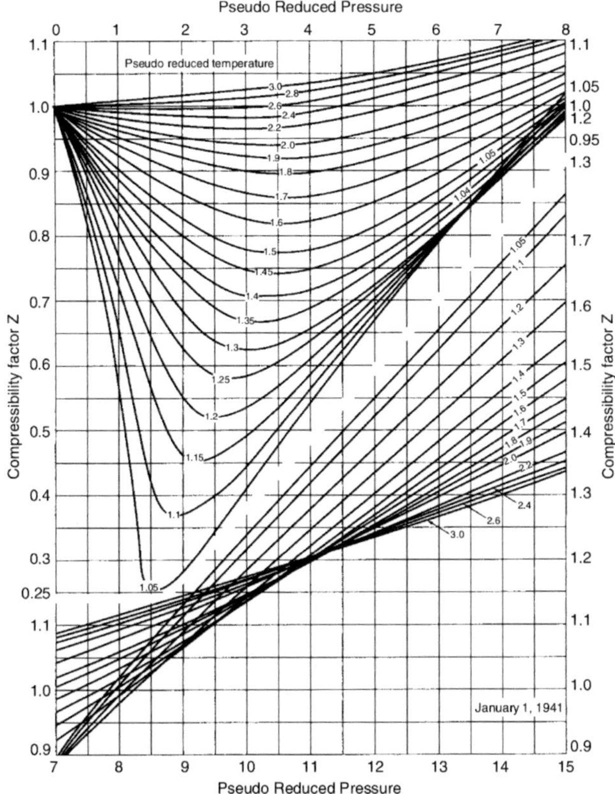

Figure 3.6. Gas compressibility factor after Katz (Lyons 2010)

The parameters a and b mixture are estimated from every individual components as well as the interaction between components according to following equations

$$a = \sum_{i=1} \sum_{j=1} y_i\, y_j \sqrt{(a_i a_j)} \cdot (1 - k_{ij}) \tag{3.50}$$

$$b = \sum_{i=1} y_i \cdot b_i \tag{3.51}$$

Where k_{ij} is the binary interaction parameters, with $k_{ii}=0$ and $k_{ij}= k_{ji}$

3.1.5. Thermal dynamic properties of mixtures

In ideal mixture, with fugacity and activity coefficient equal to one, thus the mole fraction ratio for pure component among different phases is equal to equilibrium constants, which is give in following equations. The equilibrium constants for water and hydrate are expressed in the terms of solubility and vapor pressure.

$$\begin{cases} \frac{X_w^\kappa}{X_g^\kappa} = K_{wg}^\kappa \\ \frac{X_h^\kappa}{X_g^\kappa} = K_{hg}^\kappa \\ \frac{X_h^\kappa}{X_w^\kappa} = K_{hw}^\kappa \end{cases}$$

(3.52)

K_{wg}^κ, K_{hg}^κ and K_{hw}^κ are the equilibrium constant of gas-water, water-hydrate and gas-hydrate mixture phase [-].

The viscosity of pure component can be expressed by a function in the term of pressure and temperature (Eq. 3.53).

$$\mu_\kappa = f(T, P)$$

(3.53)

In this work, the specific viscosity of gas is calculated according to the function proposed by Irvine and Liley (1984). The hydrate viscosity is computed by using a corresponding state method (Reid et al. 1987). For water and water vapor, its viscosity is estimated from the IAPWS formulation (Cooper and Dooley 1994)

The viscosity of gas mixture and hydrate mixture is expressed in Eq. 3.54 and Eq. 3.55 respectively.

$$\mu_g = \sum_{\kappa=1}^{N} \frac{X_g^\kappa \mu_\kappa}{\sum_{\lambda=1}^{N} X_g^\lambda \Phi_{\kappa\lambda}}$$

(3.54)

$$\mu_h = \prod_\kappa \mu_\kappa X_h^\kappa$$

(3.55)

Where $\Phi_{\kappa\lambda}$ is the binary interaction parameters.

The fluid enthalpy is an important property for non-isothermal flow. The specific enthalpy of fluid is composed of internal energy and the external work to change volume. In ideal gas, the internal energy is integrated from reference temperature under constant volume according to Eq. 3.56.

$$h_T = \int_{T_0}^{T} C_p^0 \, dT$$

(3.56)

Where C_p^0 is the heat capacity of ideal gas [J/mol/K]:

The enthalpy due to volume change is enthalpy departure, which is integrated from reference pressure under constant temperature according to Eq. 3.57.

$$h_p = \int_{p_0}^{p} \left(\frac{\partial h}{\partial p}\right) dp \tag{3.57}$$

The energy consumption in vaporization process is expressed as vaporization enthalpy in following

$$\Delta h_v = h^v - h^l \tag{3.58}$$

The enthalpy is a state function relating to temperature and pressure. It is only dependent on the beginning state and end state, no matter what the integral path. Hydrate phase is treated kinds of fluid. Their integral path is shown in Fig. 3,7 with beginning state at reference point. Therefore, the enthalpy for gas, water and hydrate can be expressed by Eq. 3.59-61 respectively.

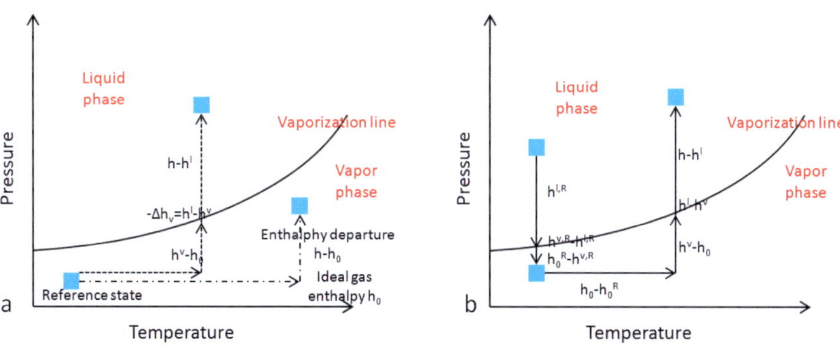

Figure 3.7. Integral path of fluid at different phase, (a) gas and liquid phase, (b) liquid phase

$$h_g = (h - h_0) + h_0 \tag{3.59}$$

$$h_w = (h - h^l) + (h^l - h^v) + (h^v - h_0) + h_0 \tag{3.60}$$

$$h_h = (h - h^l) + (h^l - h^v) + (h^v - h_0) + (h_0 - h_0^R) + (h_0^R - h^{v,R}) + (h^{v,R} - h^{l,R}) + h^{l,R} \tag{3.61}$$

The enthalpy of mixture in each phase is accumulated on pure enthalpy according to mole fraction-weighted factor in Eq. 3.62.

$$h_\beta = \sum_{\kappa=1}^{N} x_\beta^\kappa h_\beta^\kappa \tag{3.62}$$

Relative permeability with a range of 0~1 is another important parameter for flux. The effective permeability in pore medium can be explained as the product of intrinsic and relative permeability. The relative permeability is closely correlated to saturation. For two phases, there are two often used relative permeability functions; one is Van Genuchten function (Van Genuchten 1980).

$$k_{rg} = \sqrt[3]{S_g}\left[1 - \left(1 - S_g\right)^{\frac{n}{n-1}}\right]^{\frac{2(n-1)}{n}} \tag{3.63}$$

$$k_{rl} = \sqrt{S_l}\left[1 - \left(1 - S_g^{\frac{n}{n-1}}\right)^{\frac{n-1}{n}}\right]^2 \tag{3.64}$$

Where n is a constant.

The other one is and Brooks Corey function (Brooks and Corey1964) expressed in following

$$k_{rg} = S_g{}^2\left[1 - \left(1 - S_g\right)^{\frac{2+\lambda}{\lambda}}\right] = (1 - S_l)^2\left[1 - S_l^{\frac{2+\lambda}{\lambda}}\right] \tag{3.65}$$

$$k_{rl} = S_l^{\frac{2+3\lambda}{\lambda}} \tag{3.66}$$

λ is also a constant.

Some example curves for Van Genuchten and Brookes Corey is plotted in Fig.3.8.

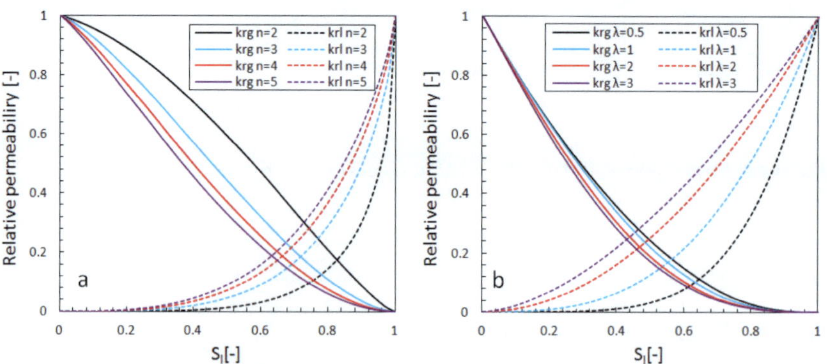

Figure 3.8. Relative permeability at different parameters after (a) van Genuchten and (b) Brooks Corey

For three phase mixture, the relative permeability function is much complicated. In this work, the modified Stone's relative permeability function (Falta et al. 1995) in terms of saturation is adopted for three phase fluid.

$$k_{rg} = \left[\frac{S_g - S_{gr}}{1 - S_{wr}}\right]^n \tag{3.67}$$

$$k_{rw} = \left[\frac{S_w - S_{wr}}{1 - S_{wr}}\right]^n \tag{3.68}$$

$$k_{rh} = \left[\frac{1 - S_g - S_w - S_{hr}}{1 - S_g - S_{wr} - S_{hr}}\right]\left[\frac{1 - S_{wr} - S_{hr}}{1 - S_w - S_{hr}}\right]\left[\frac{(1 - S_g - S_{wr} - S_{hr})(1 - S_w)}{(1 - S_{wr})}\right]^n \tag{3.69}$$

Likewise, there are also two popular capillary pressure functions in two-phase mixture, which are van Genuchten capillary pressure (Van Genuchten 1980) and Brooks Corey capillary pressure (Brooks and Corey 1964). They are given in Eq.3.70 and Eq.3.71, respectively.

$$p_{cap}(S_w) = \frac{1}{\alpha}\left(S_e^{-\frac{1}{m}} - 1\right)^{\frac{1}{n}} \tag{3.70}$$

$$p_{cap}(S_w) = p_d S_e^{-\frac{1}{\lambda}} \tag{3.71}$$

Where S_e is the effective saturation

Some example curves for capillary pressure according to van Genuchten and Brooks Corey function is plotted in Fig. 3.9 respectively.

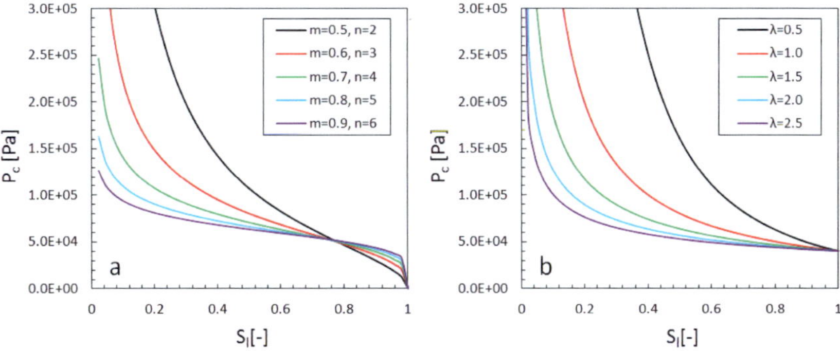

Figure 3.9. Capillary pressure at different parameters after (a) van Genuchten and (b) Brooks Corey

The capillary pressure function proposed by Parker et al. (1987) is applied to evaluate the capillary pressure for three phases. Gas-hydrate capillary pressure and gas-water capillary pressure are given in Eq. 3.72 and Eq. 3.73 individually.

$$p_{cgh} = -\frac{\rho_w g}{\alpha_{gh}}\left[(\bar{S}_l)^{\frac{n}{1-n}} - 1\right]^{\frac{1}{n}} \tag{3.72}$$

$$p_{cgw} = -\frac{\rho_w g}{\alpha_{hw}}\left[(\bar{S}_w)^{\frac{n}{1-n}} - 1\right]^{\frac{1}{n}} - \frac{\rho_w g}{\alpha_{gh}}\left[(\bar{S}_l)^{\frac{n}{1-n}} - 1\right]^{\frac{1}{n}} \tag{3.73}$$

Here $\bar{S}_w = (S_w - S_m)/(1 - S_m)$, $\bar{S}_l = (S_w + S_h - S_m)/(1 - S_m)$

3.2. Mechanic response

In this thesis, the basic mechanical behavior due to pressure change is simulated by popular program FLAC3D, which is a commercial code based on finite difference method (Itasca 2011). The governing equation and numerical method will be briefly introduced in this section. The detailed damage and fracture mechanics will be explained in following chapter.

3.2.1. Governing equations

As shown in Fig. 3.10, total three normal stresses (σ_{xx}, σ_{yy}, σ_{zz}) and six shear stresses (τ_{xy}, τ_{yx}, τ_{yz}, τ_{zy}, τ_{xz}, τ_{zx}) are used to describe the 3D stress state at a point in the Cartesian coordinate system. Because of symmetry of shear stress ($\tau_{xy}=\tau_{yx}$, $\tau_{yz}=\tau_{zy}$, $\tau_{xz}=\tau_{zx}$), the number of stress component can be reduced to six. Based on a unite element, the dynamic equilibrium equations in x-, y- and z-directions can be written in Eq. 3.74.

$$\begin{cases} \dfrac{\partial \sigma_{xx}}{\partial x} + \dfrac{\partial \tau_{xy}}{\partial y} + \dfrac{\partial \tau_{xz}}{\partial z} + F_x - \rho\dfrac{\partial^2 u_x}{\partial t^2} = 0 \\[2mm] \dfrac{\partial \tau_{xy}}{\partial x} + \dfrac{\partial \sigma_{yy}}{\partial y} + \dfrac{\partial \tau_{yz}}{\partial z} + F_y - \rho\dfrac{\partial^2 u_y}{\partial t^2} = 0 \\[2mm] \dfrac{\partial \tau_{xz}}{\partial x} + \dfrac{\partial \tau_{yz}}{\partial y} + \dfrac{\partial \sigma_{zz}}{\partial z} + F_z - \rho\dfrac{\partial^2 u_z}{\partial t^2} = 0 \end{cases} \tag{3.74}$$

Here F_x is the direction specific body force [N/m³]; u_x is the direction specific displacement [m] (see in Fig. 3.10); ρ is the mass density [kg/m³]

These equilibrium equations explain the relationship between external forces and stress under dynamic state. Set $\frac{\partial^2 u_x}{\partial t^2} = 0$, the equilibrium equations for static state is achieved.

Another important equation groups, i.e. geothermic equations are introduced to describe the correlation between displacement and strains. Correspondingly, total six strains (three normal strain ε_{xx}, ε_{yy}, ε_{zz} and three shear strain ε_{xy}, ε_{yz}, ε_{xz}) are enough to exhibit a 3D strain state in Cartesian coordinate system. The geometric equations are given in following

$$\begin{cases} \varepsilon_{xx} = \dfrac{\partial u_x}{\partial x}, & \varepsilon_{xy} = \dfrac{1}{2}\left(\dfrac{\partial u_x}{\partial y} + \dfrac{\partial u_y}{\partial x}\right) \\[2mm] \varepsilon_{yy} = \dfrac{\partial u_y}{\partial y}, & \varepsilon_{yz} = \dfrac{1}{2}\left(\dfrac{\partial u_z}{\partial y} + \dfrac{\partial u_y}{\partial z}\right) \\[2mm] \varepsilon_{zz} = \dfrac{\partial u_z}{\partial z}, & \varepsilon_{xz} = \dfrac{1}{2}\left(\dfrac{\partial u_z}{\partial x} + \dfrac{\partial u_x}{\partial z}\right) \end{cases} \tag{3.75}$$

In equilibrium equations and geometric equations, there are total fifteen unknowns, including six stresses, six strains and three displacements. Apparently, only equilibrium equations and geometrics equations is not enough to solve these unknowns. Thus, here six constitutive equations (Hooke's law) based on elastic theory (Eq. 3.76) are introduced to describe the relation between stresses and strains.

$$\begin{cases} \varepsilon_{xx} = \dfrac{1}{E}\left[\sigma_{xx} - \upsilon\left(\sigma_{yy} + \sigma_{zz}\right)\right], & \varepsilon_{xy} = \dfrac{\tau_{xy}}{2G} \\[2mm] \varepsilon_{yy} = \dfrac{1}{E}\left[\sigma_{yy} - \upsilon\left(\sigma_{xx} + \sigma_{zz}\right)\right], & \varepsilon_{yz} = \dfrac{\tau_{yz}}{2G} \\[2mm] \varepsilon_{zz} = \dfrac{1}{E}\left[\sigma_{zz} - \upsilon\left(\sigma_{xx} + \sigma_{yy}\right)\right], & \varepsilon_{zx} = \dfrac{\tau_{zx}}{2G} \end{cases} \tag{3.76}$$

Where E is Young's modulus [Pa]; υ is the Poisson ratio [-]; G is the shear modulus [Pa].

On base of these equations, any force can be solved at a given deformation, or any deformation can be solved under known external forces. The mentioned equation can be simplified in following equations according to Einstein notation. The Einstein notation is also adopted in this section at subscripts, $i,j \in (x,y,z)$.

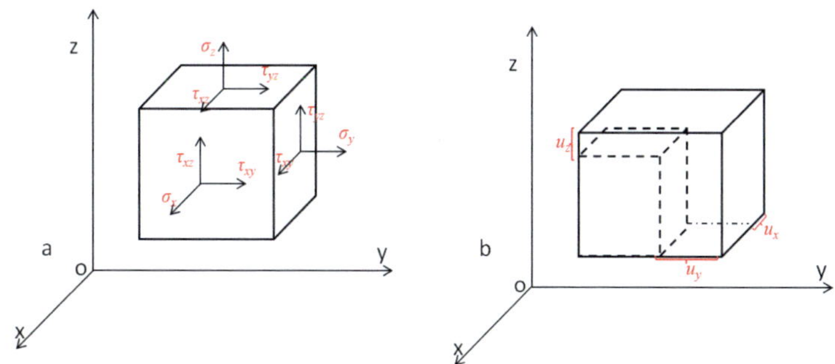

Figure 3.10. Notation of (a) stress element and (b) displacement

$$\sigma_{ij,j} + F_i - \rho\frac{dv_i}{dt} = 0 \tag{3.77}$$

$$\Delta\varepsilon_{ij} = \frac{1}{2}\left(\Delta u_{i,j} + \Delta u_{j,i}\right) \tag{3.78}$$

$$\Delta\boldsymbol{\sigma} = \boldsymbol{D}\Delta\boldsymbol{\varepsilon} \tag{3.79}$$

Where v_i is the velocity in specific direction [m/s]; $\Delta\boldsymbol{\sigma}$ is the stress increment tensor; $\Delta\boldsymbol{\varepsilon}$ is the strain increment tensor; $\boldsymbol{D}$ is the physical matrix.

3.2.2. Space discretization

For energy conservation, the internal work rate equals to external work rate for any virtual velocity increment. Hence, on the integral domain Ω shown in Fig. 3.11, we must have

$$\int_{\Omega}\boldsymbol{\sigma}:\Delta\dot{\boldsymbol{\varepsilon}}\,d\Omega = \int_{\Gamma_f}\boldsymbol{f}.\Delta\boldsymbol{v}\,d\Gamma_f + \int_{\Omega}\rho\left(\boldsymbol{b} - \frac{dv}{dt}\right)\Delta\boldsymbol{v}\,d\Omega \tag{3.80}$$

Here $\int_{\Omega}\boldsymbol{\sigma}:\Delta\dot{\boldsymbol{\varepsilon}}\,d\Omega$ is the integral form of internal work rate; $\int_{\Gamma_f}\boldsymbol{f}.\Delta\boldsymbol{v}\,d\Gamma_f$ is the integral form of external force work rate; $\int_{\Omega}\rho\left(\boldsymbol{b} - \frac{dv}{dt}\right)\Delta\boldsymbol{v}\,d\Omega$ is the integral form of external work rate due to velocity change. $\boldsymbol{f}$ is the external force [N]; Γ_f is the integral boundary of force; Γ_u is the integral boundary of displacement; $\Delta\dot{\boldsymbol{\varepsilon}}$ is strain increment rate tensor, which is defined as

$$\Delta\dot{\varepsilon}_{ij} = \frac{1}{2}\left(v_{i,j} + v_{j,i}\right) \tag{3.81}$$

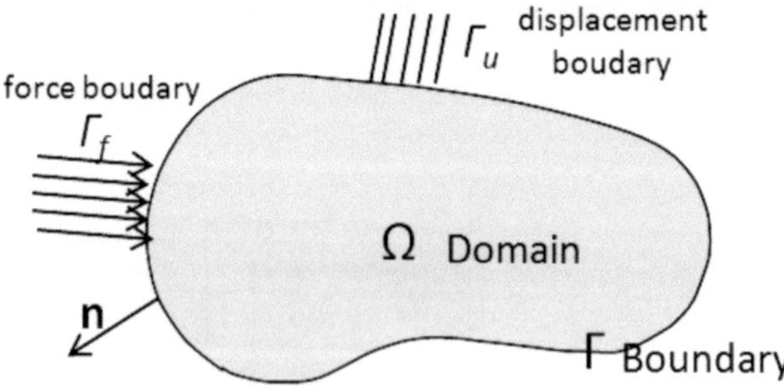

Figure 3.11. Notation of integral domain

In numerical formulation, the space derivative is approximated by using the finite difference method. In FLAC3D, the discrete element will be divided into sub tetrahedron element as in Fig. 3.12. On a tetrahedron element, the velocity divergence inside a closed surface can be replaced by the integral of outward flux on the closed surface, according Gauss divergence theory.

$$\int_{\Omega}v_{i,j}\,d\Omega = \int_{\Gamma_u}v_i\cdot n_j\,d\Gamma_u \tag{3.82}$$

Where n_j is the exterior normal direction in unite.

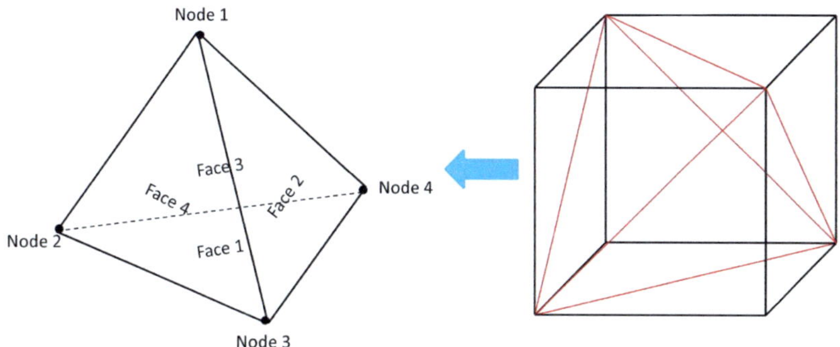

Figure 3.12. Notation of the nodes and faces of a tetrahedron element

The velocity divergence of a tetrahedron element can be in the terms of node velocity v_i^l.

$$v_{i,j} = \frac{1}{3V}\sum_{f=1}^{4}\sum_{l=1,l\neq f}^{4} v_i^l n_j^{(f)} S^{(f)} = -\frac{1}{3V}\sum_{l=1}^{4} v_i^l n_j^{(l)} S^{(l)} \tag{3.83}$$

Where f, l are the face and node index, respectively; $S^{(l)}$ is the area of surface l; $n_j^{(l)}$ is the exterior normal direction in unite; $\bar{v}_i^{(f)} = \frac{1}{3}\sum_{l=1,l\neq f}^{4} v_i^l$, means that the velocity of a surface is averaged form the three touched grid points.

Replace the strain increment rate by Eq. 3.81 and introduce the velocity divergence in Eq. 3.83, the internal work rate in a tetrahedron element can be expressed as

$$\int_{\Omega} \boldsymbol{\sigma}:\Delta\dot{\boldsymbol{\varepsilon}}\, d\Omega = \int_{\Omega} \sigma_{ij}\frac{1}{2}\left(v_{i,j} + v_{j,i}\right)d\Omega = \int_{\Omega} \sigma_{ij}v_{i,j}d\Omega = -\frac{1}{3}\sum_{l=1}^{4}\Delta v_i^l T_i^l \tag{3.84}$$

Where $T_i^l = \sigma_{ij}n_j^{(l)}S^{(l)}$

The work rate for external force f_i^l at node is written as

$$\int_{\Gamma_f} f_i\Delta v_i d\Gamma_f = \sum_{l=1}^{4}\Delta v_i^l f_i^l \tag{3.85}$$

Similarly, on the base of finite difference method, the external work rate contributions of body force and velocity change on a tetrahedron element are expressed in Eq. 3.86 and Eq. 3.87 respectively.

$$\int_\Omega \Delta v_i \rho b_i d\Omega = \sum_{l=1}^{4} \Delta v_i^l \frac{\rho b_i V}{4} = \sum_{l=1}^{4} \Delta v_i^l \, m^l b_i \qquad (3.86)$$

$$\int_\Omega \Delta v_i \rho \frac{dv}{dt} d\Omega = \sum_{l=1}^{4} \Delta v_i^l \frac{\rho V}{4} \left(\frac{dv_i}{dt}\right)^l = \sum_{l=1}^{4} \Delta v_i^l \, m^l \left(\frac{dv_i}{dt}\right)^l \qquad (3.87)$$

Where $m^l = \frac{\rho V}{4}$, V is the volume of tetrahedron element.

Substituting Eqs. 3.84-87 into Eq. 3.82, and summing all the contributing tetrahedron and node, the general form of work equilibrium at l-th node can be written as

$$m^l \left(\frac{dv_i}{dt}\right)^l = \frac{T_i^l}{3} + m^l b_i + f_i^l = F_i^l \qquad (3.88)$$

Here F_i^l is the sum of global contribution at node l.

Any small disturbance or unbalance force will change the movement of grid point, further change the deformation energy. Under the effect of friction, such movement will slow down and finally stop, at which time a static state is achieved. To reach static or quasi-static state quickly, a damping force, which has a direction opposite to the movement, is introduced in Eq. 3.88.

$$m^l \left(\frac{dv_i}{dt}\right)^l = F_i^l + Fd_i^l \qquad (3.89)$$

Where Fd_i^l is the damping force [N], which is defined as $Fd_i^l = -\alpha |F_i^l| sign(v_i^l)$

$$sign(y) = \begin{cases} +1 & if\ y > 0 \\ -1 & if\ y < 0 \\ 0 & if\ y = 0 \end{cases} \qquad (3.90)$$

Solve Eq. 3.88 using the explicit finite difference method, the velocity of node l at new time step is calculated according to following equation.

$$v_i^l(t + \Delta t) = v_i^l(t) + \frac{\Delta t}{m^l}\left(F_i^l(t) + Fd_i^l(t)\right) \qquad (3.91)$$

On the base of the new node velocity as well as time step (Δt), the node displacement increment at new time ($t + \Delta t$) can be achieved, which will be further used to evaluate the new stress and strain increment at node with the help of Eq. 3.78 and Eq. 3.79 respectively.

3.2.3. Plastic deformation and failure criterions

Some typical strain-stress curves are shown in Fig.3.13. When the stress state still in elastic region, it shows a linear correlation between stresses and strains (see Eq. 3.76). If the loading is removed, the deformation can recover fully. However, once stress state reaches or exceeds the yield point, it enters plastic region with a non-linear correlation between stresses and strains. The deformation in plastic region is irreversible. The strain-stress curves of different type of rocks behaviors different after yield point, including strain-hardening, ideal elastic-plastic, strain softening as well brittle. These behaviors are classified according to the rest strength.

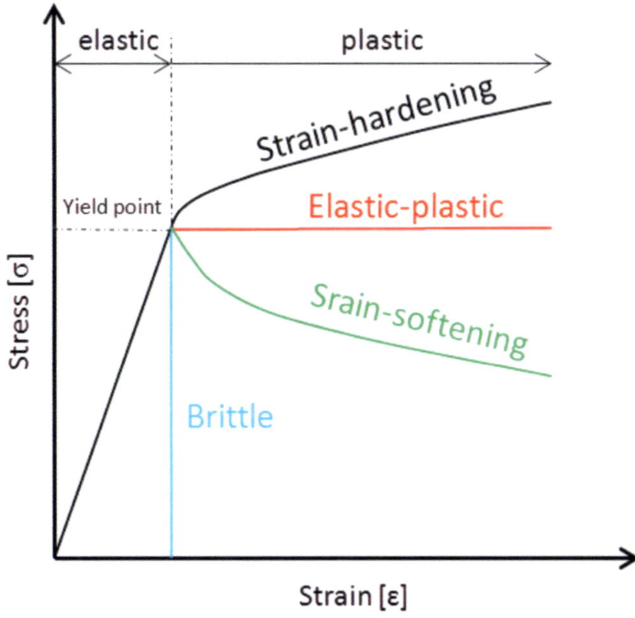

Figure 3.13. Typical plastic strain-stress curves

The boundary of elastic and plastic deformation is defined by yield point, namely failure criterion, which is strongly dependent on the stress state. There are two very popular failure criterions. One is Mohr failure criterion, in which the failure happens for the reason that shear stress over the shear strength. The expression of Mohr failure criterion is written in Eq. 3.93 or Eq. 3.94.

$$F^P = \tau - \sigma_n \cdot \tan\varphi - c \tag{3.92}$$

or

$$F^P = \sigma_1 - \sigma_3 \cdot N_\varphi - 2c\sqrt{N_\varphi} \tag{3.93}$$

Where σ_n is the effective normal stress [Pa]; φ is the friction angle [°]; c is the cohesion [Pa]; σ_1, σ_3 are maximum and minimum principal stress respectively [Pa]; N_φ is a constant defined as

$$N_\varphi = \frac{1+sin\varphi}{1-sin\varphi}$$

Another failure criterion is tensile criterion (Eq. 3.94). In this criterion, the plastic failure takes places, when the tensile stress exceeds tensile strength.

$$F^P = -\sigma_3 - \sigma_T \tag{3.94}$$

Where σ_T is tensile strength [Pa].

To explain the relationship between stress state and failure criterion, the stress state represented by Mohr circles is drawn in Fig. 3.14 together with Mohr and tensile failure criterion line. When Mohr circle locate at the right side of failure criterion, the stress state is located elastic. If the Mohr cricle reaches or crosses the failure lines, then plastic deformation will be triggered. From Fig. 3.14, it can be found that tensile failure is triggered more likely with a smaller Mohr circle, namely lower difference between maximum and minimum effective principal stress.

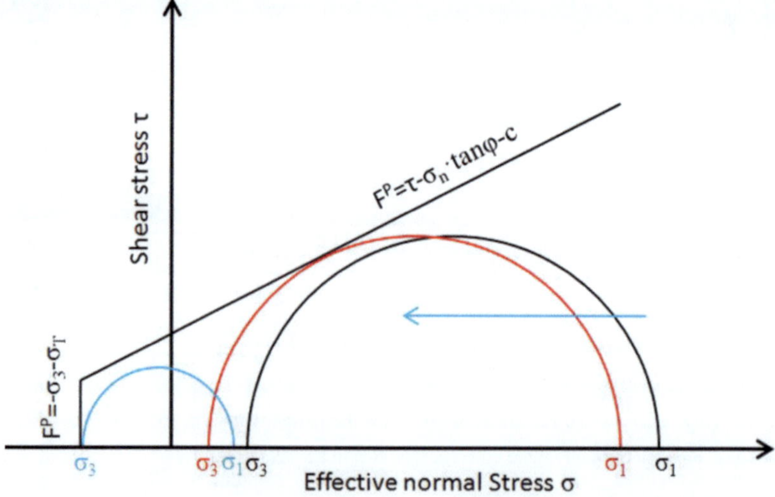

Figure 3.14. Demonstration of the relationship between Mohr circle and failure criterion

The plastic deformation can be calculated by following equation

$$d\varepsilon_{ij}^P = d\lambda \cdot \frac{\langle F^P \rangle}{|F^P|} \cdot \frac{\partial Q^P}{\partial \sigma_{ij}} \qquad (3.95)$$

Where ε_{ij}^P is plastic strain [-]; λ is plastic strain multiple [-]; Q^P is potential function; F^P is failure function; $\langle \, \rangle$ operator is defined as following

$$\langle F^P \rangle = \begin{cases} 0 & if \ F^P \leq 0 \\ F^P & if \ F^P > 0 \end{cases}$$

The potential functions corresponding to Mohr and tensile failure criterion are given in Eq. 3.96 and Eq. 3.97, respectively.

$$Q^P = \sigma_1 - \sigma_3 \cdot N_\varphi \qquad (3.96)$$

$$Q^P = -\sigma_3 \qquad (3.97)$$

3.3. THM coupling processes

The multi-components and -flow as well mechanic behavior of reservoir have been introduced in previous section independently. However, in the real geo-energy reservoirs, these processes, mainly categorized in thermal, hydraulic and mechanic process, run tightly coupled with each other. The coupling concept as well as implemented schema will be discussed in this section.

3.3.1. Coupling concept

The THM coupled process is very complicated in geo-energy reservoir. The coupling effect among these processes is illustrated in Fig. 3.15. The effect of temperature on fluid can be characterized by the thermal dynamic properties of fluid, like viscosity and density, etc. Inversely, the impact flow on thermal process is taken into account by the flow-dependent heat advection and the departure enthalpy relating to hydraulic pressure. The coupling interactions between hydraulic and mechanic processes are described by effective stress, porosity and permeability. The effect of pressure is brought into mechanic by effective stress. On the contrast, the effective stress-dependent porosity and permeability can influence the hydraulic process directly. Due to the thermal expansion property, temperature plays a very important role on the mechanic behavior of rock. However, mechanic process affects thermal process by shear heating due to friction, which is too weak to be considered in this work.

To quantify the effect of thermal as well as hydraulic process on mechanic, thermal stress and pressure acting on porous rock are defined as

$$\sigma^{thr} = -3\beta K \Delta T \tag{3.98}$$

$$\sigma^{hyr} = \alpha p \tag{3.99}$$

Where σ^{thr} is the thermal stress [Pa]; β is the thermal expansion coefficient [1/°C]; K is the bulk modulus of grain [Pa]; ΔT is the temperature increment [°C]; σ^{hyr} is the partial pressure acting on porous rock [Pa]; α is the biot coefficient [-]; p is the pore pressure [Pa].

Substituting Eq. 3.98 and 3.99 into Eq. 3.76, then

$$\boldsymbol{\sigma} - (\alpha P + 3\beta K \Delta T)\boldsymbol{I} = \lambda tr \boldsymbol{\varepsilon} \boldsymbol{I} + 2\mu \boldsymbol{\varepsilon} \tag{3.100}$$

Here $\boldsymbol{I}$ is the unite matrix; $\boldsymbol{\sigma}$ is the stress tensor; $\boldsymbol{\varepsilon}$ is the strain tensor; μ is the shear modulus [Pa]; λ is the Lame first parameter [Pa]; tr is the tensor operator of trace.

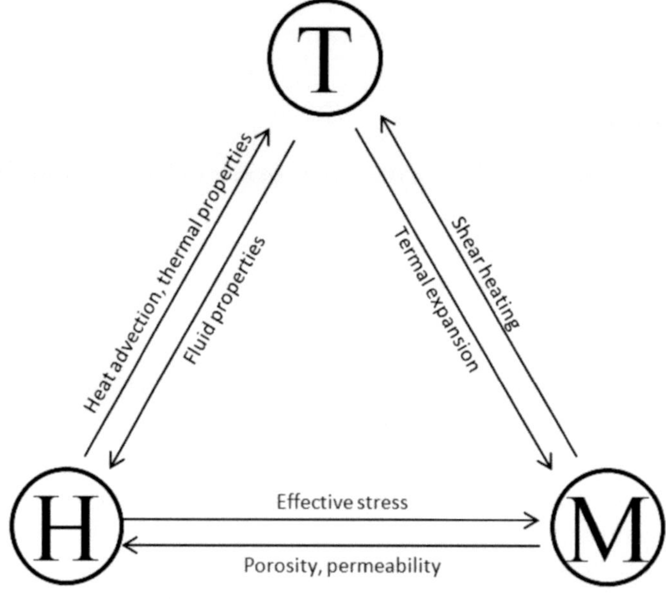

Figure 3.15. THM coupled process

Referring to Rutqvist and Tsang (2002), the formulas of stress-dependent porosity and permeability can be written in Eq. 3.101 and 3.102, whose original formulas were derived on the base data over

1000 direct laboratory measurement of permeability in core plug samples at various conditions of net effective stress (Davies and Davies 1999).

$$\phi = (\phi_0 - \phi_r)exp(5 \times 10^{-8} \cdot \sigma'_m) + \phi_r \tag{3.101}$$

$$k = k_0 \left[22.2 \left(\frac{\phi}{\phi_0} - 1\right)\right] \tag{3.102}$$

Where ϕ_0 is the porosity at zero stress condition [-]; ϕ_r is the residual porosity [-]; $\sigma'_M = \sigma_M - \alpha p$, is the effective mean stress [Pa]; σ_m is the mean stress [Pa].

3.3.2. Data flow in THM modeling

As shown in Fig. 3.16, the data need be exchanged between hydraulic thermal codes THOUGH2MP and mechanic code FLAC3D to achieve THM coupling. Obviously, the precondition is that the discrete elements in two programs need one-to-one correspondence. Besides, the geometry of model and discrete element need identical for a high computing precious. In each loop, the mean stress averaged at the element center is gathered firstly from all elements and sent to THOUGH2M through the platform of WinStock. After receiving the data from FLAC3D, the data will be distributed to individual processors on corresponding element for porosity and permeability update. In THOUGH2MP, the temperature and pressure at new time step is calculated according to multi-flow introduced in previous section. The primary parameters, like temperature and pressure, are gathered again from all elements in separate processors in specific order. After that, the gathered temperature and pressure are sent back, further distributed to corresponding elements in FLAC3D. The received temperature and pressure are applied in Eq. 3.99 to get stress redistribution for next time step. This is the data flow with one THM coupled cycle and it will continue until reaching the time limit.

The data exchange introduced here is only one cycle for the fundamental THM coupled model. In numerical model specific for geothermal energy and unconventional gas, the data exchange is much complicated and will be introduced in following chapters.

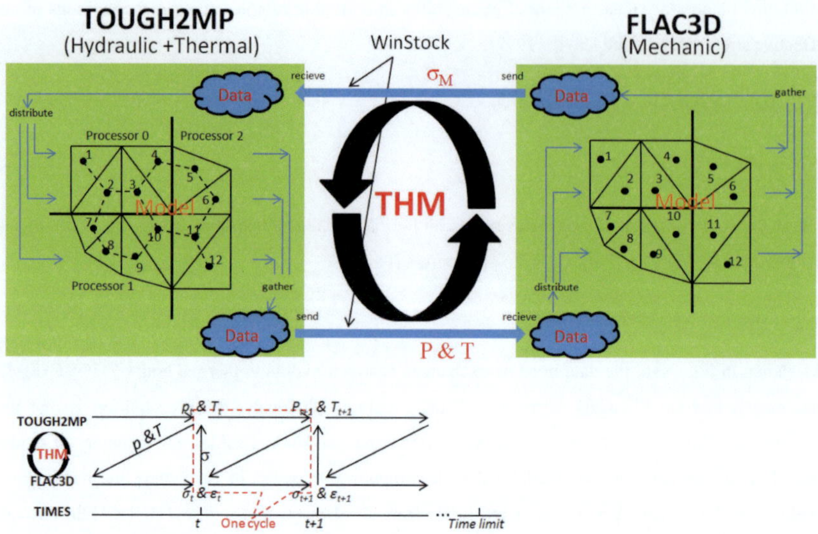

Figure 3.16. Data flow in THM coupled framework

4. CO_2 fracturing in unconventional gas reservoirs

4.1. CO_2 fracturing concepts

As a useful method, hydraulic fracturing has been widely used to enhance the gas recovery from unconventional reservoir, which owns a low even ultra-low permeability (Economides and Nolte 2000). The most commonly used is water-based fracturing fluid, due to its lower price and excellent performance. However, massively hydraulic fracturing could bring some serious problem, like underground water pollution, water shortage. In addition to this, some water-sensitive clay mineral limits the application of water-based hydraulic fracturing. To overcome these problems, an alternative fracturing fluid-CO_2 has been proposed and attracts extensive attention in the last two decades. In previously numerical as well as experimental studies (Zhang et al. 2017; Wang et al. 2018), CO_2-based fracturing fluid performs very well in low break down pressure, creating a more complex fracture-network as well as more gas recovery through the desorbed CH_4 by CO_2. Besides, re-utilization of CO_2 in unconventional reservoir could partly reduce the greenhouse emission. However, before widely application, there remain some barriers need be removed, like high leakage caused by low viscosity and poor proppant-carrying ability.

As the CO_2 property is very temperature- and pressure-sensitive, the CO_2 fracturing simulator need be integrated in the THM coupled framework FLAC3D-THOUGH2MP. To characterize the fracture propagation, a potential plane fracture elements is generated perpendicular to the minimum principal stress and intersected in the middle of discrete reservoir element, since the fracture propagation direction is assumed perpendicular to the minimum principal stress. As shown in Fig. 4.1, the fracture opening, closing as well as propagating only take place within the fracture potential path. The corresponding fracture models as well as proppant transport model are developed on this potential fracture path. Moreover, the multi-flow model in fracture has been derived on the variable fracture width. The multi-fluid exchanged between fractured elements with its host reservoir elements, is treated as fluid leakage. Three additional components, including surfactant thickened CO_2, polydimethylsiloxane (PMDMS) thickened CO_2 and guar gum are defined in this section, to compare their performance in unconventional reservoir. The thermal impact of CO_2 fracturing is discussed in this chapter as well. Then, the different combination of thickened CO_2 and proppant in different densities has been compared with traditional fracturing fluid-guar gum, to address the weak proppant-carrying ability of CO_2.

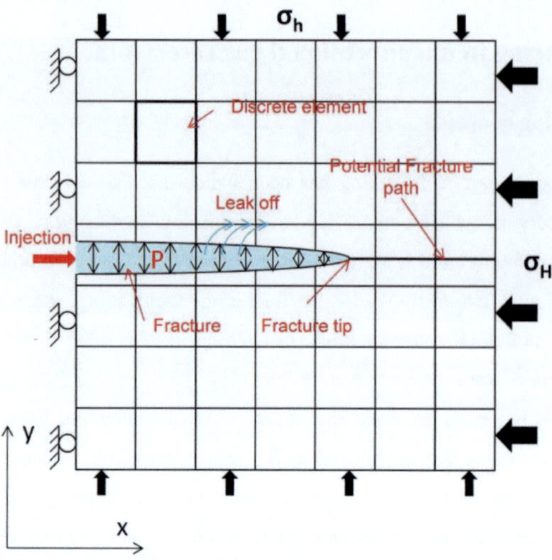

Figure 4.1. Demonstration of the fracture element in discrete model

4.2. Description of relevant models

Multi-flow model:

As introduced above, discrete element is divided into two types, reservoir element and potential fracture element. For component κ, the fluid motion in both discrete elements is characterized by Darcy law with a formula in following

$$\vec{u}_\beta = -k\frac{k_{r\beta}\rho_\beta}{\mu_\beta}\left(\vec{\nabla}P_\beta - \rho_{m\beta}\vec{g}\right) \tag{4.1}$$

Where $\vec{u}_\beta$ is the fluid velocity in phase β (gas, aqueous and hydrate) [m/s]; $k_{r\beta}$ is the relative permeability in phase β [m²]; ρ_β is the mol density in phase β [mol/m³]; $\rho_{m\beta}$ is the mass density in phase β [kg/m³]; μ_β is the viscosity in phase β [Pa·s]; P is the pressure [Pa]; g is the gravity [N/m]; k is the permeability [m²].

The averaged reservoir permeability is used to calculate the fluid motion between reservoir elements. When the mobile fluid flows from fracture to reservoir or from reservoir to fracture, the permeability of the host reservoir element will be adopted to describe the flow. But for the flow in fractures, the permeability between parallel fracture walls is approximated by following equations (Zhou 2014; Liao et al. 2020a).

$$k = \frac{(fw)^2}{12} \tag{4.2}$$

Here f is the factor defines fracture roughness [-].

According to Li et al. (2018), the mass conservation in discrete reservoir element and potential fracture element are given in Eq. 4.3 and Eq. 4.4, respectively.

$$\frac{\partial\left(\phi \sum_\beta S_\beta \rho_\beta x_\beta^\kappa\right)}{\partial t} = -\vec{\nabla}\left(\sum_\beta \vec{u}_\beta x_\beta^\kappa\right) + q^\kappa \tag{4.3}$$

$$F * \frac{1}{w} * \frac{\partial\left(w \sum_\beta S_\beta \rho_\beta x_\beta^\kappa\right)}{\partial t} = -\vec{\nabla}\left(\sum_\beta \vec{u}_\beta x_\beta^\kappa\right) + q^\kappa \tag{4.4}$$

Here ϕ is the porosity [-]; F is the fraction of fractured element in total element, which is defined in following section [-]. w is the fracture width [m]; q^k is the injection source of component κ [m³/s].

These mass conservation equations are solved by using Newton iteration with the help of code TOUGH2MP. As introduced in section 3, corresponding accumulation M^k and residuum term R_n^{t+1} of component κ in arbitrary reservoir elements is given in following

$$M^\kappa = \sum_\beta \phi S_\beta \rho_\beta x_\beta^\kappa \tag{4.5}$$

$$R_n^{\kappa,t+1} = M_n^{\kappa,t+1} - M_n^{\kappa,t} - \frac{\Delta t}{V_n} \sum_m FA_{mn} u_{mn}^{\kappa,t+1} - \Delta t q_n^{\kappa,t+1} = 0 \tag{4.6}$$

Here V_n is the volume of control element [m³]; u_{nm}^{t+1} is the fluid velocity [m/s]; Δt is the time step [s]; A_{mn} is the contact area between two elements [m²], the contact area between two fracture element is dependent on the fracture width and factor F.

For the accumulation and residuum in potential fracture element can be given in following

$$M^\kappa = F \sum_\beta w S_\beta \rho_\beta x_\beta^\kappa \tag{4.7}$$

$$R_n^{\kappa,t+1} = M_n^{\kappa,t+1} - M_n^{\kappa,t} - \frac{\Delta t w^{t+1}}{V_n} \sum_m A_{mn} u_{mn}^{\kappa,t+1} - \Delta t w^{t+1} q_n^{\kappa,t+1} = 0 \tag{4.8}$$

After solving these equations, the primary parameters, including pressure, temperature and fraction of components in formation as well as in fracture are gained at new time step.

Fracture model:

As the fracture pressure acts directly on fracture wall in the host reservoir element, the fracture width increment can be evaluated according to the current pressure in fracture and the geometry of its host reservoir element.

$$\Delta w^{t+1} = \frac{\Delta \sigma_n}{\alpha_1} l_c = \frac{P^{t+1} - \sigma_n^t}{\alpha_1} l_c \tag{4.9}$$

Here $\Delta \sigma_n$ is the stress increment [Pa]; P^{t+1} is the current pressure [Pa]; lc is the width of host reservoir element [m]; α_1 is the rock toughness [Pa], $\alpha_1 = K + 4G/3$; $\alpha_2 = K - 2G/3$.

In Eq. 4.9, if current fracture pressure exceeds the compression stress in host reservoir element, the fracture width will be enlarged. Conversely, if fracture pressure is lower, the fracture width will be reduced. However, fracture width cannot reduce infinitely, because proppant as well as fracture roughness will support the fracture, when the fracture wall contacts with proppant. Therefore, a contact stress introduced by Zhou (2014) is used to evaluate the fracture width increment in following

$$\Delta w^{t+1} = \frac{\Delta \sigma_n - \sigma_{con}^t}{\alpha_1} l_c = \frac{P^{t+1} - \sigma_{con}^t - \sigma_n^t}{\alpha_1} l_c \tag{4.10}$$

Where σ_{con} is the contact stress [Pa] with

$$\sigma_{con}^{t+1} = \begin{cases} 0 & if\ Cp \leq 0.65\ or\ w > w_{res} \\ \sigma_{con}^t + \alpha_1 \Delta \varepsilon_0 & if\ Cp > 0.65\ or\ w \leq w_{res} \end{cases}$$

$\Delta \varepsilon_0$ is the over reduced strain [-]

Due to the fracture width increment, the stress state on corresponding host reservoir element will be altered according to following equations

$$\sigma_n^{trial,t+1} = \sigma_n^t + \alpha_1 \frac{\Delta w}{l_c} \tag{4.11}$$

$$\sigma_{2,3}^{trial,t+1} = \sigma_{2,3}^t + \alpha_2 \frac{\Delta w}{l_c} \tag{4.12}$$

Mechanical model:

The trial stress is applied firstly on corresponding host reservoir element. Such trial stress will break the stress equilibrium state. Then, with the new pore pressure in reservoir element, the equilibrium state can be achieved again by solving following equations (Itasca 2011).

$$\sigma_{ij,j} + \rho g_i = \rho \frac{\partial v_i}{\partial t} \tag{4.13}$$

$$\Delta\varepsilon_{ij} = \frac{1}{2}\Delta t\left(\frac{\partial v_i}{\partial x_j} + \frac{\partial v_j}{\partial x_i}\right) \tag{4.14}$$

$$\boldsymbol{\sigma'} = \boldsymbol{D\varepsilon} \tag{4.15}$$

where $\sigma = \sigma' - \alpha IP + 3\beta K\Delta T$, v_i is velocity [m/s]; $\Delta\boldsymbol{\varepsilon}$ is strain tensor [-]; $\Delta\boldsymbol{\sigma'}$ is effective stress tensor [Pa]; α is Biot's-coefficient [-]; β is thermal expansion coefficient [1/°C]; K is bulk modulus [Pa]; I is unit matrix; and $\boldsymbol{D}$ is physical matrix; $i; j \in (1,2,3)$, E is Young's modulus [Pa]; v is Poisson's ratio [-].

$$D = \frac{E}{(1+v)(1+2v)}\begin{bmatrix} 1-v & v & v & & & \\ v & 1-v & v & & & \\ v & v & 1-v & & & \\ & & & 1-2v & & \\ & & & & 1-2v & \\ & & & & & 1-2v \end{bmatrix}$$

Fracture propagation:

As shown in Fig. 4.2, the potential fracture path is inserted into the middle of reservoir element, in which the potential fracture element is divided into three types, fractured, partial fractured and non-fractured element. In order to improve the accuracy, the fracture tip will be divided into sub-element. On every sub-element, the tensile failure criterion (Eq. 4.16) will be checked on the base of current stress and pressure. If enough number of sub-elements is fractured, the fracture tip will be converted into fractured element. After that, non-fractured elements surrounding the new fractured element are assigned as new fracture tip again, for failure criterion check in next time step. Then the fraction of fractured element is estimated by Eq. 4.17.

$$P^{t+1} - \sigma_n^{t+1} > \sigma_T \tag{4.16}$$

$$F = \frac{N_{fractured}}{N_{tot}} \tag{4.17}$$

Where σ_T is tensile strength [Pa]; $N_{fractrrured}$ is the number of fractured sub-element [-]; N_{tot} is total number of sub-element [-].

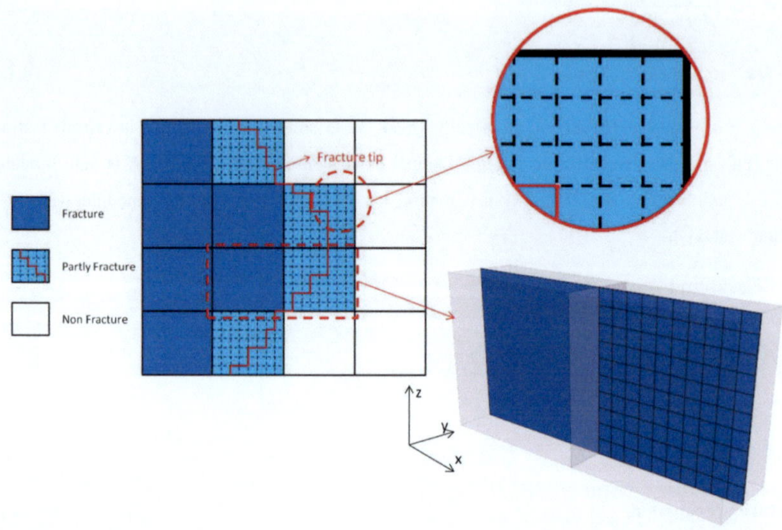

Figure 4.2. Illustration of the fractured, partly fractured and non-fractured element and sub-element (Liao et al. 2019a)

Proppant transport model:

Zhou (2014) and Hou et al. (2015a) introduced a proppant transport model. In this model, the slurry velocity (u_s) shows a linear correlation with proppant velocity (u_p) and liquid velocity (u_l), which is given in following

$$u_s = (1 - Cp)u_l + Cpu_p \tag{4.18}$$

Here u_s is the slurry velocity [m/s]; u_l is the liquid velocity [m/s]; u_p is the proppant transport velocity [m/s]; Cp is the proppant concentration [-].

The proppant transport is driven mainly by liquid velocity as well as gravity. Physically, the part velocity driven by liquid is not exactly equal to liquid velocity. Referring to Liu (2006), a relative velocity model is introduced to characterize correlation between liquid velocity and proppant velocity driven by fluid, in which the influence of proppant diameter, fracture width as well as proppant concentration are considered.

$$k_{wc} = \left. u_{pl} \middle/ u_l \right. = \begin{cases} -0.826\left(\frac{d_p}{w_{eff}}\right)^2 + 0.0475\left(\frac{d_p}{w_{eff}}\right) + 1.2713 & if \ \left(\frac{d_p}{w_{eff}}\right) < 0.93 \\ -7.71\left(\frac{d_p}{w_{eff}}\right) + 7.71 & if \ \left(\frac{d_p}{w_{eff}}\right) \geq 0.93 \end{cases} \tag{4.19}$$

Where d_p is the proppant diameter [m]; w_{eff} is the effective fracture width [m], which is defined as in the formula of fracture width, proppant diameter and proppant concentration.

$$\frac{1}{w_{eff}} = \frac{1}{w^2} + 1.411\left(\frac{1}{d_p{}^2} - \frac{1}{w^2}\right)Cp^{0.8}$$

Generally, the density of proppant is higher than liquid, because it is difficult to transport low denser proppant down to perforation, if buoyancy is larger than gravity. Therefore, proppant particle will settle down in slurry. The corresponding terminal settling velocity of a single proppant particle can be described in following equation, which relates to the particle Neynolds number.

$$u_{pst} = \begin{cases} \dfrac{d_p{}^2(\rho_p - \rho_l)g}{18\mu_a} & N_{Rep} < 2 \\ \dfrac{0.79132d_p{}^{1.14}(\rho_p - \rho_l)^{0.71}}{\rho_l{}^{0.29}\mu_a{}^{0.43}} & N_{Rep} \geq 2 \end{cases} \tag{4.20}$$

ρ_p is the proppant density [kg/m³]; ρ_l is the liquid density [kg/m³]; μ_a is the apparent viscosity [Pa·s], N_{Rep} is the particle Neynolds number, which is given in following

$$N_{Rep} = \frac{d_p u_p \rho_l}{\mu_a}$$

Two additionally correlation factors are introduced to describe the effects of proppant concentration as well as wall hindering according to the correlation gained in experimental results (Barre and Conway 1994).

$$c_{corr} = 2.37Cp^2 - 3.08Cp + 1 \tag{4.21}$$

$$w_{corr} = 0.563\left(\frac{d_p}{w_{eff}}\right)^2 - 1.563\left(\frac{d_p}{w_{eff}}\right) + 1 \tag{4.22}$$

Where c_{corr} is the correlation of proppant concentration [-]; w_{corr} is the correlation of wall hindering effect [-].

Then proppant velocity can be expressed in Eq. 4.23 by the combined velocity driven by liquid and gravity.

$$u_p = k_{wc}u_l + u_{pst}c_{corr}w_{corr}cos\theta = k_{wc}u_l + u_{ps}cos\theta \tag{4.23}$$

Where θ is the angle between flux and vertical direction [°]; u_{sp} is the proppant settling velocity [m/s].

Substituting Eq. 4.23 into Eq. 4.18, the liquid and proppant velocity can be expressed further by Eqs. 4.24-25 respectively.

$$u_l = \frac{u_s - Cpu_{ps}cos\theta}{1+k_{wc}Cp-Cp} \tag{4.24}$$

$$u_p = \frac{u_s-(1-Cp)u_l}{Cp} \tag{4.25}$$

Then estimated proppant velocity is brought into the proppant mass conservation (Eq. 4.26) to obtain the proppant distribution.

$$\frac{\partial(wCp)}{\partial t} + \nabla \cdot \left(u_p wCp\right) + Cp_{inj}q_s \tag{4.26}$$

Fracturing fluid model

In this section, the reservoir stimulation with different fracturing fluids, involving pure CO_2, thickened CO_2 and guar gum, are performed in tight gas reservoir, thus, the fracturing fluid, guar gum and thickened CO_2 need be defined firstly. As viscosity property plays a very critical role in fracture creation, the application of different additives, like guar and CO_2 thickener attempts to improve the viscosity of base solution. Therefore, the viscosity property of pure CO_2, thickened CO_2 and guar gum is characterized in following part. Regarding to other properties, like compressibility and enthalpy is calculated according to the base fluid of solution, like water and CO_2, which is already given in Pruess and Battistelli (2002).

As introduced in Torres et al. (2014), the relationship between apparent viscosity and zero-shear-rate viscosity for guar gum can be written in

$$\mu = \frac{\mu_0}{\left(1+k\gamma^{(1-n)}\right)} \tag{4.27}$$

Where μ is the apparent viscosity [Pa·s]; μ_0 is the zero-shear-rate viscosity [Pa·s]; k is the time-constant [-]; n is the flow index [-].

Through matching the data from Torres et al. (2014), the guar concentration (C_g)-dependent time-constant and flow index is expressed in flowing

$$k(C_g) = a_2 * b_2{}^{ln(C_g)} \tag{4.28}$$

$$n(C_g) = a_3 * ln(C_g) + b_3 \tag{4.29}$$

Here a_2, b_2, a_3, b_3 is matching constant [-]; C_g is the guar concentration [g/L].

Besides, the zero-shear-rate viscosity is also guar concentration dependent. Their correlation is given in following

$$\mu_0(C_g) = a_1 * b_1{}^{ln(C_g)} \qquad (4.30)$$

a_1, b_1 are also matching constant [-].

Referring to Zhang et al. (2007), the Arrhenius equations is introduced to describe the thermal influence on the zero-shear-rate viscosity.

$$\mu_0(C_g, T) = a_1 * b_1{}^{ln(C_g)} exp\left(\frac{E_a}{R}\left(\frac{1}{T} - \frac{1}{T_0}\right)\right) \qquad (4.31)$$

Where E_a is the activation energy [J/mol]; R is the ideal gas constant [J/mol/K]; T_0 is the reference temperature [K], here is set as 20+273.15K.

According to above equations, the apparent viscosity, predicted under different guar concentrations and shear rates, are compared in Fig. 4.3(a), at a constant room temperature. All used parameters are summarized in tab. 4.1. The predicted apparent viscosity shows a good agreement with the original data from Torres et al. (2014). To study the thermal effect, the zero-shear-rate viscosity of guar gum at concentration of 10g/L is predicted under different temperature and compared with the data from Zhang et al. (2007). The zero-shear-rate viscosity decreases, but the decreasing rate is gradually slow down with an increasing temperature. As shown in Fig. 4.3b, the predicted zero-shear-rate viscosity matches very well with the data from Zhang et al. (2007).

Table 4.1. Value of the parameters

Parameter	Value	Parameter	Value
a_1	2.6431×10^{-3}	b_1	26.9127
a_2	4.1080×10^{-3}	b_2	9.2018
a_3	2.5900×10^{-3}	b_3	2.3210×10^{-1}
E_a	19515.70	R	8.314

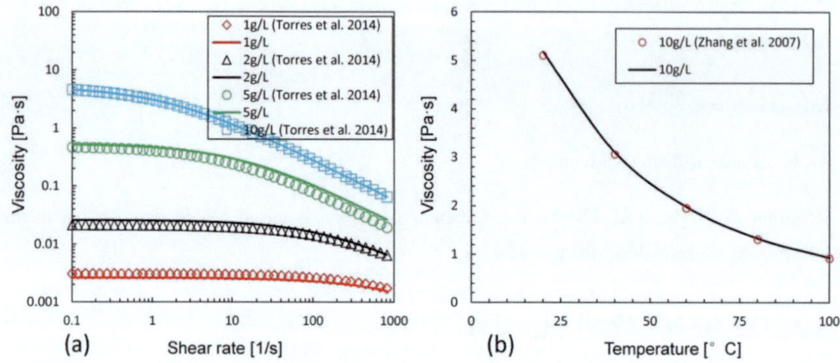

Figure 4.3. (a) Comparing the predicted apparent viscosity with data from Torres et al. (2014), (b) Comparing the predicted zero-shear-rate viscosity with data from Zhang et al. (2007) under different temperature

As shown in Fig. 4.4 (a), the apparent viscosity of guar gum with a concentration of 5g/L is predicted under different temperatures as well as shear rates. Clearly, the apparent viscosity is falls down with growing shear rate. And a very similar tendency is observed under different temperatures. At the same shear rate, the apparent viscosity shows an inverse ratio with temperature. Additionally, the apparent viscosity under different shear rates and concentration are plotted in Fig. 4.3(b). The apparent viscosity of guar gum increases dramatically, when the guar concentration increases from zero to about 5g/L. Then the growing rate slows gradually with further increasing guar concentration. Besides, at low shear rates, the variation apparent viscosity due to shear rate is slight in comparison with high shear rate.

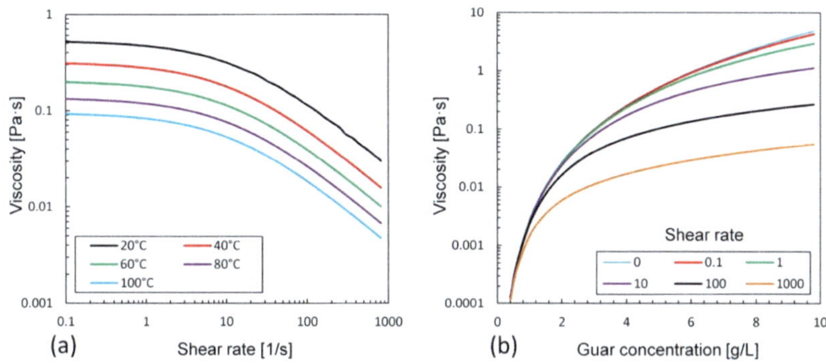

Figure 4.4. (a) Apparent viscosity versus shear rate with guar concentration of 5g/L under different temperatures, (b) Apparent viscosity versus guar concentration under different shear rate

Apparently, the proppant in fracture, particularly high proppant concentration, could influence the fluid flow in fracture. According to Zhou (2014), the impact of proppant concentration on apparent viscosity can be accounted by the help of following equation

$$\mu(C_g, T, C_P) = \mu(C_g, T)\left(1 - \frac{C_p}{C_{p,max}}\right)^{-a} \tag{4.32}$$

where C_p, max is the maximum proppant concentration, a is the correlation coefficient.

Set C_p, max =0.65 and $a = 1.8$, the relationship between apparent viscosity and proppant concentration is plotted in Fig. 4.5, in which guar concentration and temperature are set as 5g/L and 10°C, respectively. Obviously, the proppant concentration in fracture will increase the apparent viscosity of guar gum in fracture.

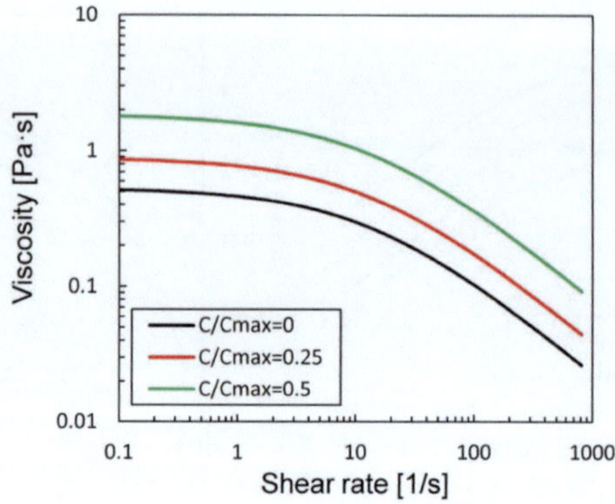

Figure 4.5. Apparent viscosity versus shear rate under different proppant concentration

Due to low viscosity causing significant leakage, the application of pure CO$_2$ as fracturing fluid is limited in unconventional gas reservoir with relatively low permeability. In response to this, the CO$_2$ thickener was suggested by many researchers for CO$_2$ fracturing (see in Tab. 2.3). In this work, two CO$_2$ thickeners, including surfactant and polydimethylsiloxane (PDMS) will be used, which are named as Thicker 1 and Thickener 2 respectively in following section. According to Luo et al. (2015) and Li et al. (2019b), the temperature influence on surfactant as well as PDMS thickened CO$_2$ viscosity can be approximated also by the Arrhenius equations,

$$\mu = A_v exp\left(\frac{E_f}{R_g T}\right)$$

(4.33)

Where μ is the viscosity [Pa·s], A_v is the pre-exponential factor [Pa·s]; E_f is the activation energy [J/mol], R_g is the molar gas constant[J/mol/K].

Table 4.2. Value of the parameters

Parameter	Pre-exponential factor(A_v)	activation energy(E_f)
Surfactant (Thickener 2)	4.25×10^{-2}	15000
PDMS (Thickener 1)	2.76×10^{-3}	15000
Pure CO$_2$	8.50×10^{-5}	15000

The viscosity of different thickened CO_2 as well as pure CO_2 are predicted by using Eq. 4.33 and plotted in Fig. 4.6. The applied parameters are summarized in Tab. 4.2. The surfactant thickened CO_2 has relatively high viscosity in the range from 20×10^{-3} to 0.5×10^{-3} Pa·s with the temperature increasing from 20°C to 120°C, which is comparable with the data from Luo et al. (2015). The viscosity of PDMS thickened CO_2 shows a similar tendency but low value in comparison with surfactant thickened CO_2. The predicted viscosity of PDMS thickened CO_2 locates dominant in the range that given by Li et al. (2019b). The lowest viscosity is found in pure CO_2, ranging about 0.07×10^{-3} to below 0.01×10^{-3} Pa·s. The viscosity of surfactant thickened CO_2 has been improved about tens to hundreds times in comparison with the one of pure CO_2.

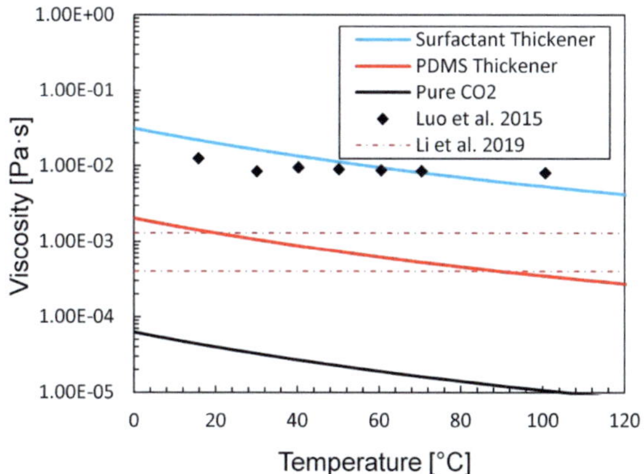

Figure 4.6. The apparent viscosity of thickened CO_2 as well as pure CO_2 under the impact of temperature

4.3. Working schema and data flow of the CO_2 fracturing model

As shown in Fig. 4.7, the simulation begins with model generation on the base of geological information. After inputting all necessary parameters, the simulation enters the first cycle. The initial parameters, like temperature, fracture pressure, fracture width and fluid density etc. is assigned into FLAC3D to achieve a new stress distribution. Based on new fracture width, fluid velocity density as well as fluid density, the proppant placement will be calculated according to proppant transport model. Meanwhile, the failure criterion in sub-elements at fracture tip will be checked to judge fracture propagation. Then the stress state, proppant concentration, contact stress as well as updated fraction of fractured sub-element in whole fracture element are sent to reservoir simulator TOUGH2MP, where the multi-flows in both fracture and reservoir elements are solved coupling with TMVOC EOS. The apparent viscosity of fracturing fluid is determined according to fracturing fluid model. In addition to this, the fracture width increment is estimated in fracture model and sent back to FLAC3D together with other primary parameters for next cycle. The simulation will continue, till the time reaches the time limit. This fracturing model is also developed on the popular THM framework TOUGH2MP-FLAC3D as introduced in section 3.3.3. The data flow in this fracturing model is illustrated in Fig. 3.8. Only the viscosity, velocity and density of fracturing fluid are exchanged between TOUGH2MP and FLAC3D, which is used to estimate the proppant placement in fracture. As introduced in section 3.3.1, the permeability as well as porosity of reservoir will be updated on the base of current stress state from FLAC3D according to Eq. 3.100-3.101.

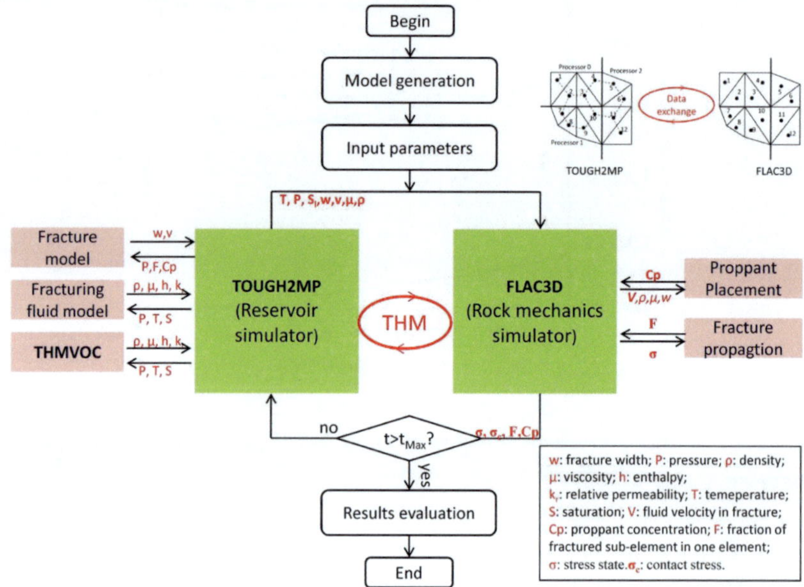

Figure 4.7. Computing schema under a THM coupled framework

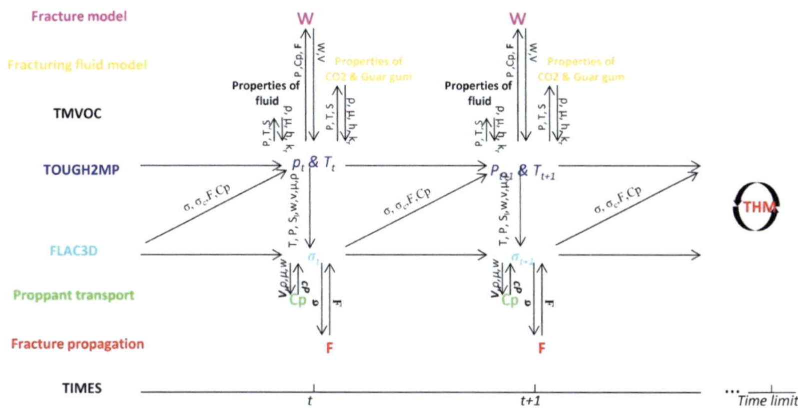

Figure 4.8. Time step based data flow under THM coupled framework

4.4. Verification of implemented numerical models

The mentioned fracturing model, involving fracture model, multi-flow model and proppant transport model, has been implemented into THM coupled framework TOUGH2MP-FLAC3D. These implemented models are verified separately in this section.

4.4.1. Verification of fracture model in FLAC3D

To verify the implemented fracture model, a pseudo 2D model is generated with dimension of 30m×60m in x- and y-directions, respectively. A unit thickness is adopted in this model. As shown in Fig. 4.9a, the fluid under different viscosity is injected at the middle of left boundary with a constant rate of 0.5 kg/s. Potential fracture element is assumed in x-direction crossing injection point, which is perpendicular to minimum stress. The compression stress of 20 MPa and 10 MPa are applied in x- and y-direction, respectively. The boundary condition is demonstrated in Fig. 4.9b. The simulation is carried out under room temperature. Other important parameters, including mechanical, hydraulic, thermal as well as coupling parameters are summarized in Tab. 4.3. As shown Fig. 4.9b, two important parameters, fracture width and fracture length are used to describe the created fracture geometry.

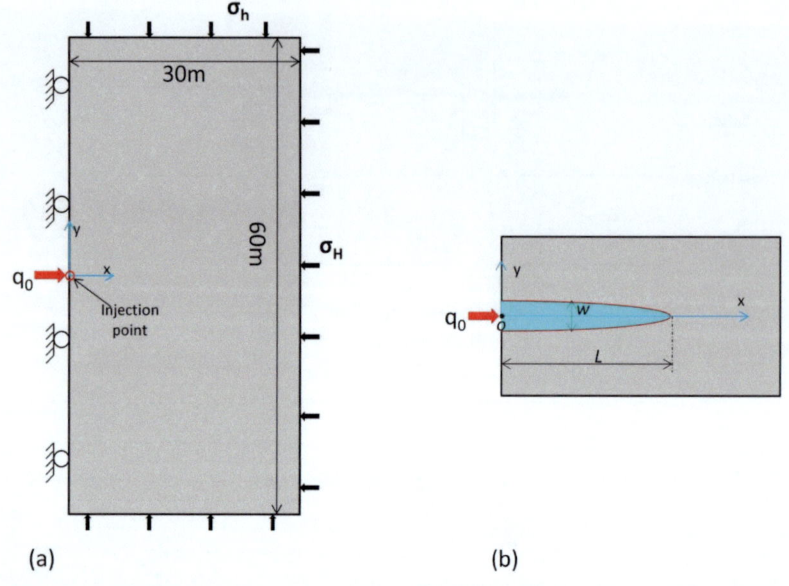

Figure 4.9. Demonstration of simple 2D model for fracture verification

Table 4.3. Summary of the used variables

Parameter	Value	Unite
Injectio rate (q)	0.5	kg/s
Young's modulus (E)	20×10^9	Pa
Poisson ratio (v)	0.25	-
Tensile strength (σ_T)	1×10^6	Pa
Permeability (k)	1×10^{-15}	m^2
Porosity (ϕ)	10	%
Thermal expansion coefficient (β)	1×10^{-6}	1/°C
Specific heat capacity (Cs)	965	J/kg/°C
Biot coefficient (α)	1	-

The simulated fracture length and fracture width are compared with the analytical solution based on classic KGD model, whose approximate formulas were suggested by Dontsov (2017) in following.

$$L(t) = 0.539 \left[\frac{q^3 E}{12\mu(1-v^2)}\right]^{1/6} t^{2/3} \tag{4.34}$$

$$w(\xi,t) = 2.36 \left[\frac{12\mu q^3(1-v^2)}{E}\right]^{1/6} t^{1/3}(1-\xi)^{0.588}(1+\xi)^{2/3} \tag{4.35}$$

Where t is the injection time [s]; v is the Poisson ratio [-]; q is the injection rate [m³/s]; μ is the fluid viscosity [Pa·s]; $\xi = \frac{x}{L}$

As shown in Fig. 4.10a, the terminal half-length under different viscose fluids match very well with analytical solutions. Apparently, high viscosity tends to create a shorter fracture, as highly viscose fluid will meet more resistance in fracture and need a higher injection pressure, which is confirmed by the fracture width profile in Fig. 4.10b. The highly viscose fluid tends to create a much wider fracture for a high net pressure. In spite of this, the simulated fracture width profile shows very good agreement with analytical solution under different fluid viscosities.

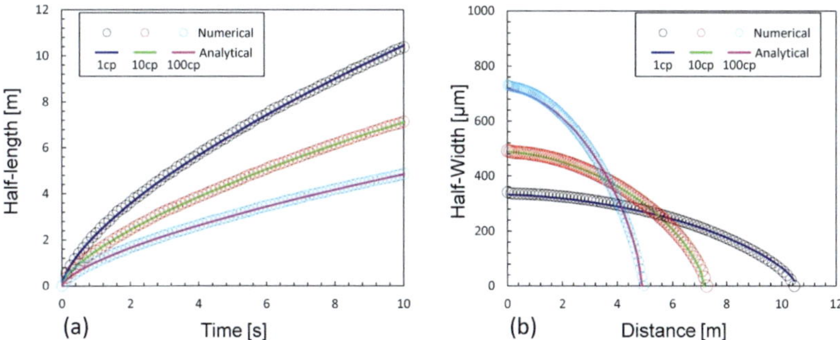

Figure 4.10. (a) Comparing the numerical and analytical fracture half length, (b) the numerical and analytical fracture half width at 10 s

4.4.2. Verification of multi-component and -flow in THOUGH2MP

The multi-component and -flow will be verified in a simple 1D model through comparing with the simulated results from Oldenburg et al. (2013). As shown in Fig. 4.11, the 1D model is 61m long and is meshed equally into 61 elements. The crossing-section is square shape with the area of 0.0929m². Pure CO$_2$ is injected at the left under a rate of 0.009 m³/s. On the other end area, the pore pressure is fixed constantly as initial pore pressure of 204bar. Rest surfaces are closed for flow. The temperature of 91.8°C and aqueous saturation of 1 is initiated in this model. Three components, CO$_2$, CH$_4$ and water are considered in this model. More detail information is given in Oldenburg et al. (2013) and Gou et al. (2016).

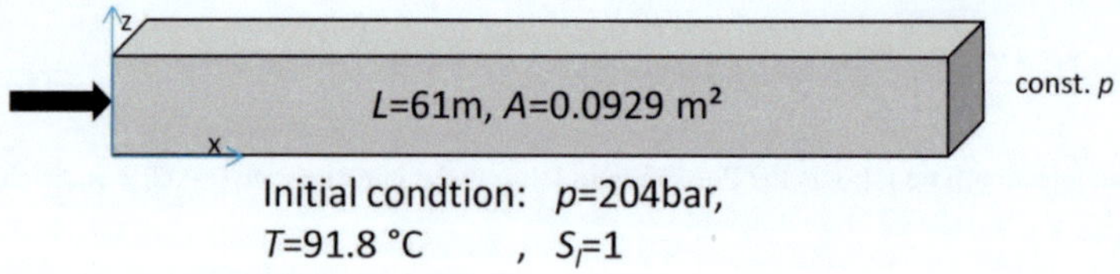

Figure 4.11. Demonstration of simple 1D model for multi-flow verification

The simulated gas pressure and saturation along the 1D model are plotted in Fig. 4.12a together with the simulated results from Oldenburg et al. (2013). The simulated pressure from present model near injection point is slightly high, but matches very well in rest part. Similarly, the simulated gas saturation in previous segment is slightly mismatching with the results from Oldenburg et al. (2013). Overall, the simulated results of present model are very comparable with the results from Oldenburg et al. (2013). The mass fractions of CO$_2$ as well as CH$_4$ in gas phase are compared in Fig. 4.12b. There is significant peak of CH$_4$ fraction at migration front of CO$_2$, because the initial CH$_4$ is concentrated by the power of injecting CO$_2$. In the previous segment, the saturation of CO$_2$ reaches up to 100%. The final results show a good agreement with Oldenburg's results. Therefore, the multi-flow in newly developed model is verified.

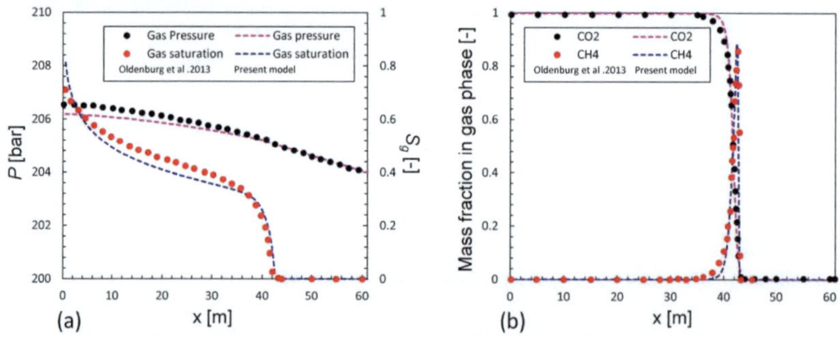

Figure 4.12. (a) Simulated gas pressure and saturation, (b) the CO$_2$ and CH$_4$ saturation along the 1D model

4.4.3. Verification of proppant transport model in THOUGH2MP-FLAC3D

To verify implemented proppant transport model, the simulated results by newly developed model is compared with the experimental results by Barree and Conway (1994), in which a 2D slot with the dimension of 4.8768m×1.2192m was constructed. A constant gap width of 7.8994mm is built in this slot. As shown in Fig. 4.13, the injection point lies at the right side about 0.3048 m below the model top, while the left side is free for out flow. Under room temperature, the guar gum containing 2.63 g/L guar is injected at a constant rate of 2.52×10^{-4} m^3/s. It contains about 479kg proppant in each cubic meter injection fluid. In this condition, the apparent viscosity of guar gum is rough 0.35 Pa·s. Other applied parameters are listed in Tab. 4.4. The simulated proppant distribution contours at 30s and 150s are illustrated in Fig. 4.14 together with the experimental results at corresponding time. The red region represents the proppant concentration larger than 0.04, which is the minimum value is calculated from the ratio between single proppant particle volume and gap volume. Obviously, the simulated proppant placement and experimental results are very comparable with each other, especially the proppant transport distance at each time point and the slope of interface between proppant and guar gum fluid at 150s. Thus, the feasibility of the proppant transport model by guar gum is verified. As the similar viscosity properties of thickened CO$_2$ and guar gum, it assumed the proppant transport model is also useful for thickened CO$_2$.

Table 4.4. Summary of the used variables

Parameter	value	unite
Injection rate (q)	2.52×10^{-4}	m^3/s
Proppant density (ρ_p)	2650	kg/m^3
Guar concentration (Cg)	2.63	g/L
Temperature (T)	25	°C
Proppant diameter (p_d)	0.6	mm
Parameter for viscosity correlation (a)	1.8	-
Maximal proppant concentration (C_{max})	0.65	-

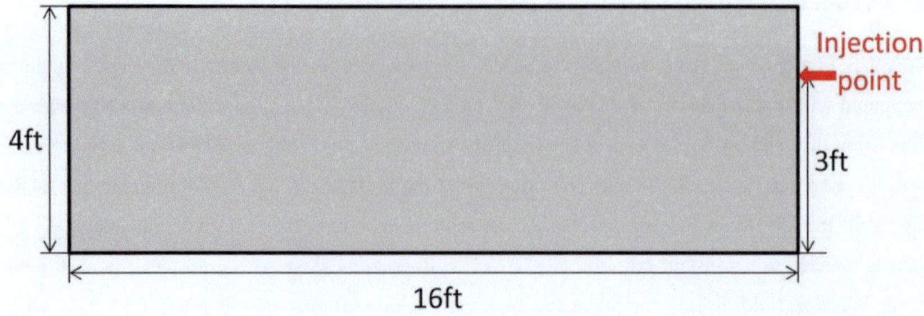

Figure 4.13. Demonstration of the 2D model for proppant transport verification

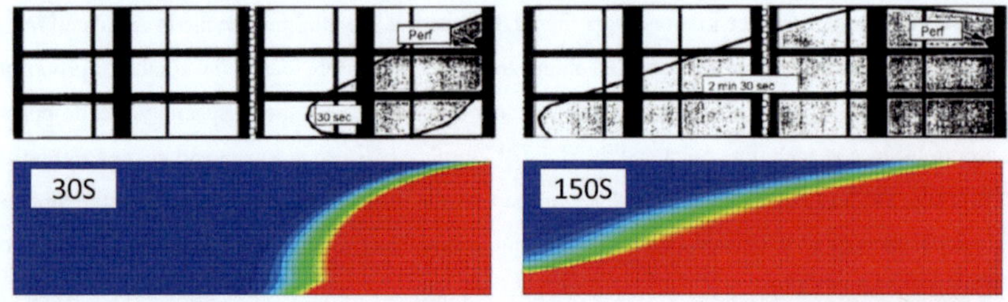

Figure 4.14. Comparing simulated proppant placement with the experiment results from Barree and Conway (1994)

4.5. A generic model of a tight gas reservoir and the initial condition

To study the performance of CO$_2$ in unconventional gas reservoir, a typical fictitious model is generated with the geometry of 200m ×300m ×200m at the depth from -3200m to -3000m. As shown in Fig. 4.15, the model is divided into three strata formation, named "basement", "Pay_zone" and "Cap_rock" from bottom to top respectively. A potential fracture plane with the same mesh is inserted in the host reservoir element at y=0. In this 1/4 part of a symmetrical model, all boundaries are closed for out flow. Yet, one point shown as red arrows is left for fluid injection, positing in the potential fracture element at the depth of -3100m. The movement on four sides plus bottom surface is fixed. Three initial stresses and pore pressure are illustrated also in Fig. 4.15 along depth. The minimum stress is found in y-direction, which is perpendicular to the potential fracture plane. A clear stress barrier is adopted in Cap_rock as well as in Basement to limit the fracture propagation in vertical direction. In addition to this, the initial temperature is converted from a common temperature gradient of 3°C/100m and surface temperature of 20 °C. Other applied parameters, including mechanical parameters, strength, hydraulic parameters and thermal parameter are listed in Tab. 4.5, wherein the permeability and porosity are 0.1mD and 10% in Pay_zone respectively. Both are relatively higher than that of Cap_rock and Basement. This is a typical tight gas reservoir.

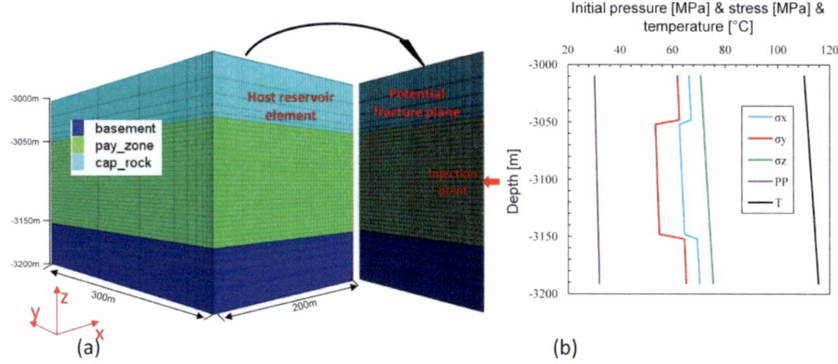

Figure 4.15. (a) Demonstration of the discrete model, (a) applied initial stress, pressure and temperature

Table 4.5. Applied mechanical, hydraulic and thermal parameters

Group	Cap_rock	Pay_zone	Basement
Young's modulus [Pa]	25×10^9	30×10^9	25×10^9
Poisson ratio [dimensionless]	0.3	0.25	0.3
Tensile strength [Pa]	2×10^6	1×10^6	2×10^6
Permeability [m²]	1×10^{-17}	1×10^{-16}	1×10^{-17}
Porosity [%]	2.5	10	2.5
Thermal expansion coefficient [1/°C]	1×10^{-6}	1×10^{-6}	1×10^{-6}
Specific heat capacity [J/kg/°C]	965	965	965
Biot coefficient [-]	1	1	1

The composition in pore of this model is demonstrated in Fig. 4.16. Only two phases are involved in the original pore. The aqueous phase, water-dominant, occupies about 20% mol weights. Rest pore is occupied by 95% CH_4 and 5% CO_2 in gas phase. As shown in Fig. 4.17, different fracturing fluids are injected at a constant rate of 6 m³/min. Finally, 480m³ fracturing fluid is injected into the whole model in 80mins. As guar gum is a guar water solution, its density is assumed equal to water of $1000kg/m^3$. However, the density of injected CO_2 is strongly related to the temperature as well as pressure. Generally, the CO_2 is pumped from a hermetic tank, in which the CO_2 is transported and stored as liquid under low temperature and high pressure. Therefore, the density of injected liquid CO_2 could higher than water. Here the liquid CO_2 with density $1150kg/m^3$ in tank is used. Physically, the cold CO_2 will be heated in the process pumping from surface to perforation underground. The influence of injection well is not involved in this numerical model. To ensure the injected CO_2 in supercritical state, the pure CO_2 as well thickened CO_2 are injected at the temperature of 40°C. In coordinate with it, the traditional fracturing fluid, water and guar gum are injected at the temperature of 40°C as well.

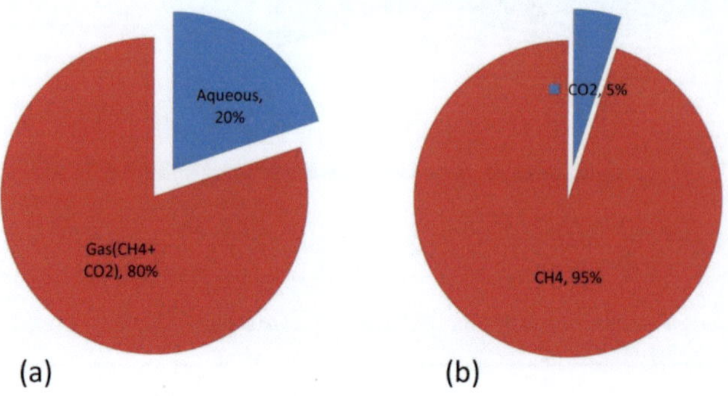

Figure 4.16. (a) Initial component in aqueous and gas phase, (b) initial proportion of CO_2 and CH_4 in gas phase

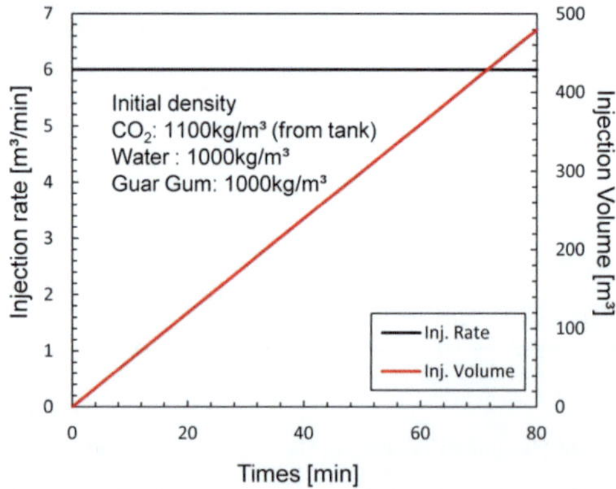

Figure 4.17. Injection plan and injection volume

4.6. Viscosity variation of the injected thickened CO_2

After 80 min injection, the simulated results of pure CO_2 and different thickened CO_2s are compared in Fig. 4.18. Obviously, the created fracture is symmetrical with respect to a horizontal line crossing injection point. Comparing the created fracture half width contours, the Thickener 2, namely surfactant thickened CO_2, performs better to create a much wider fracture, due to low leak-off caused by low viscosity. The maximum fracture half width around 1.05cm is found at injection point. Other than this, the maximum half length created by Thickener 2 could reach up to 130m, which is much longer than the other two fractures created by Thickener 1 and pure CO_2. The smallest fracture, with half length and half width only 31 m and 0.18 cm respectively, is stimulated by pure CO_2 injection. Based on these, it can be concluded that the viscosity of CO_2 plays a critical role on fracture creation in such tight gas reservoir with permeability of 0.1 mD, because large leakage makes the pure CO_2 inefficient to create a fracture for production. But if improve the viscosity of pure CO_2 by using thickener, it is possible to create an efficient fracture. The pore pressure in host reservoir element is illustrated in Fig. 4.18. Basically, the maximum value of pore pressure shows a negative correlation with fracturing fluid viscosity. In spite of this, the pore pressure due to pure CO_2 injection distributes evenly, except for it due to Thickener 2 injection, in which the high pore pressure distributes in two bands at the top and bottom of fracture around a semicircle region. The low pressure around injection point is caused by low leak-off rate due to high viscosity determined by relatively low temperature. Although the viscosity is higher at fracture tip, it has not enough time for leak-off, thus the pore pressure close to fracture is still relatively low.

In the fracture created by Thickener 2 injection, the temperature and viscosity distribution are illustrated in Fig. 4.19. During fluid flowing from injection point into fracture as well as reservoir, the cold thickened CO_2 will be heated gradually through heat exchanging with reservoir rock as well as mixing with warmer fluid in fracture. Therefore, the temperature increases gradually with the distance away from injection point, where the lowest temperature is measured equal to the injection temperature of 40 °C. The temperature in non-fractured area is almost unaffected. Especially, the temperature drop area is smaller than fracture contour, because the temperature of cold thickened CO_2 is already heated, when it reaches the fracture tip. Inversely, the apparent viscosity decreases 1.34×10^{-3} Pa·s gradually with the distance away from injection point under the influence of temperature, which confirms again the pore pressure distribution in Fig. 4.18. Besides, the pressure, temperature and viscosity along the dotted line in Fig. 4.19a are plotted at different times in Fig. 4.19d-e respectively. Obviously, the variation the pressure is in fracture is ignorable, while the pore pressure is in the shape of inverse "U" with high value in middle and low in two ends. From Fig. 4.19e, temperature drop area propagates much slowly than fracture propagating. But, the propagation of temperature drop area is consistent with viscosity change.

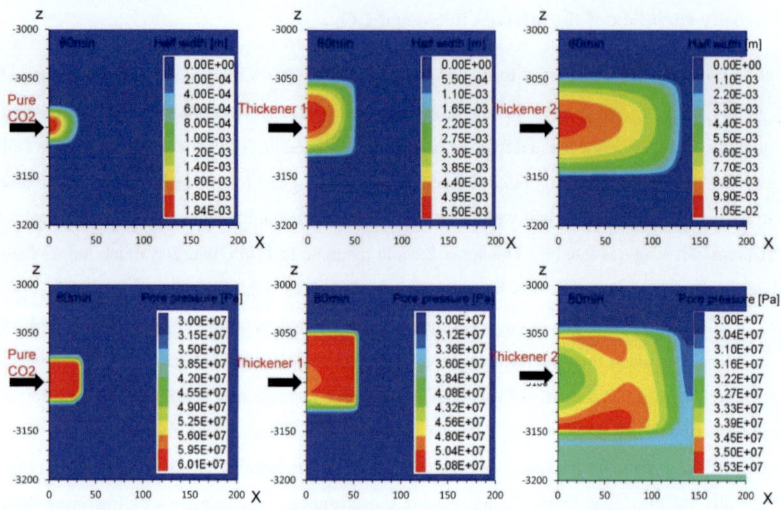

Figure 4.18. The results the fracture half width and pore pressure with different fracturing fluids

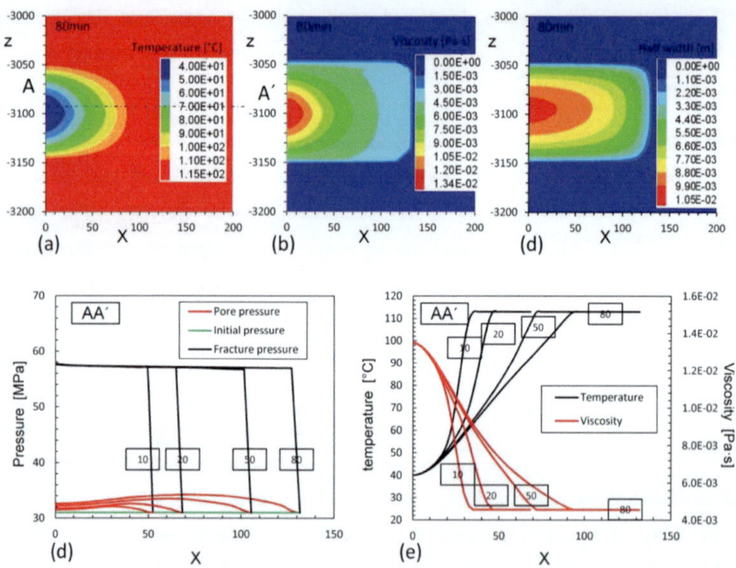

Figure 4.19. Results by Thickener 2, (a) temperature contour, (b) viscosity contour, (c) fracture half with contour, (d) fracture pressure and pore pressure at different time points, (d) temperature and viscosity along fracture at different time points

During fracturing, the injected CO_2 as well as thickened CO_2 will leak into reservoir formation to change its composition. In the fracture and reservoir formation with Thickener 2 injection, the CO_2 fraction in gas phase is shown in Fig. 4.20a-b respectively. Evidently, whole fracture is occupied by the CO_2 in gas phase. The distribution of CO_2 in reservoir formation is consistent with pore pressure in Fig. 4.18. The maximum CO_2 fraction is improved from 5% up to 36.9% in the host reservoir element. In reservoir formation, the CO_2 fraction in gas phase by different fluids injection is plotted along dotted line B-B` and C-C`, respectively. Dotted line B-B` is in host reservoir element parallel to fracture wall, while C-C` is perpendicular to fracture wall crossing the injection point. By pure CO_2 injection, fast all the CH_4 in the host reservoir element containing fracture is replaced by CO_2, which is evidenced by the CO_2 fraction close to 100% in Fig. 4.20c. However, by Thickener 1 and Thickener 2 injections, the maximum CO_2 fraction drops to about 80% and 30%, respectively. More information about CO_2 invade depth at injection point is illustrated in Fig. 4.20d. Reasonably, the invade depth shows a negative correlation with fluid viscosity and it reaches 7m, 3m and 1m by pure CO_2, Thickener 1 and 2 injection respectively.

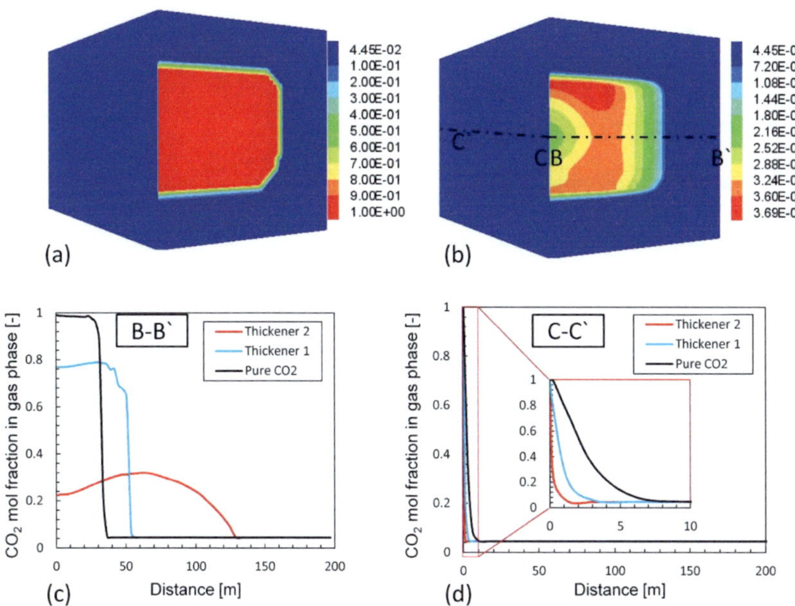

Figure 4.20. (a) CO_2 saturation in gas phase in fracture, (a) CO_2 saturation in gas phase in formation pore, (c) CO_2 saturation in gas phase along the dotted line B-B`, (a) $CO_2$ saturation in gas phase along the dotted line C-C`

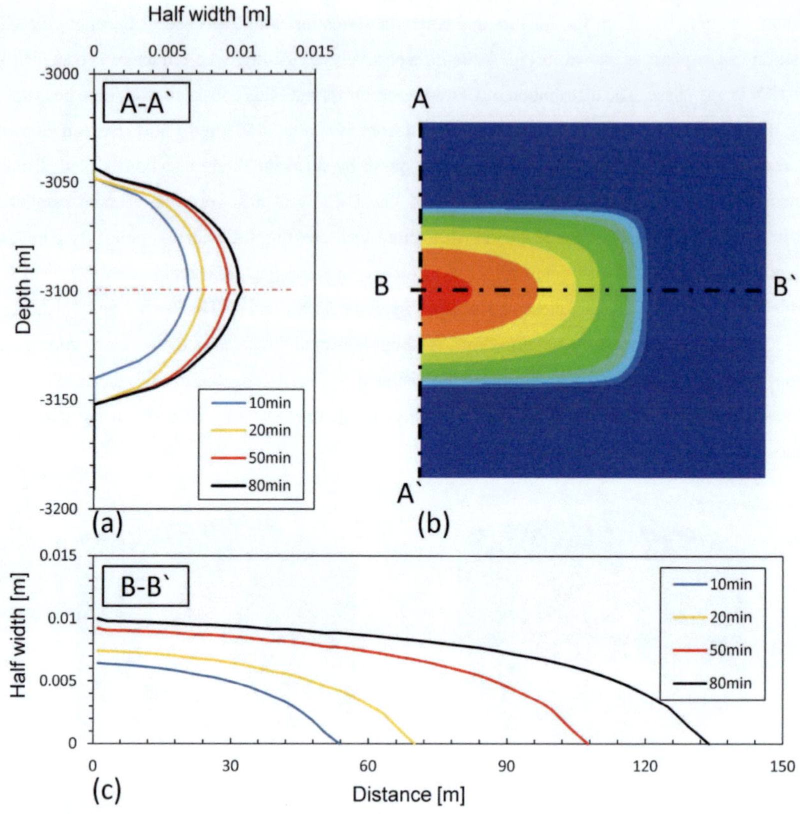

Figure 4.21. (a) Fracture width profile at different time along the dotted line A-A`, (b) fracture half with contour at 80min by Thickener 2 injection, (c) fracture profile at different time along the dotted line B-B`

At different time points, the fracture width profiles by Thickener 2 injection are shown in Fig. 4.21. As shown by the vertical fracture profile in Fig. 4.21a, the fracture half width is enlarged but with gradually decreased expanding rate. Under the effect of stress barrier, the fracture propagation in vertical is limited within 100m height after about 20 min. Similarly, the fracture propagation in horizontal direction with a gradually decreasing speed (see in Fig. 4.21c), because enhanced fracture will increases the area for leak off. Similar profiles by pure CO$_2$, Thickener 1 and 2 injections are illustrated in Fig. 4.22. The fracture created by pure CO$_2$ is too narrow to place proppant, while Thickener 2 could create enough fracture for proppant transport and placement.

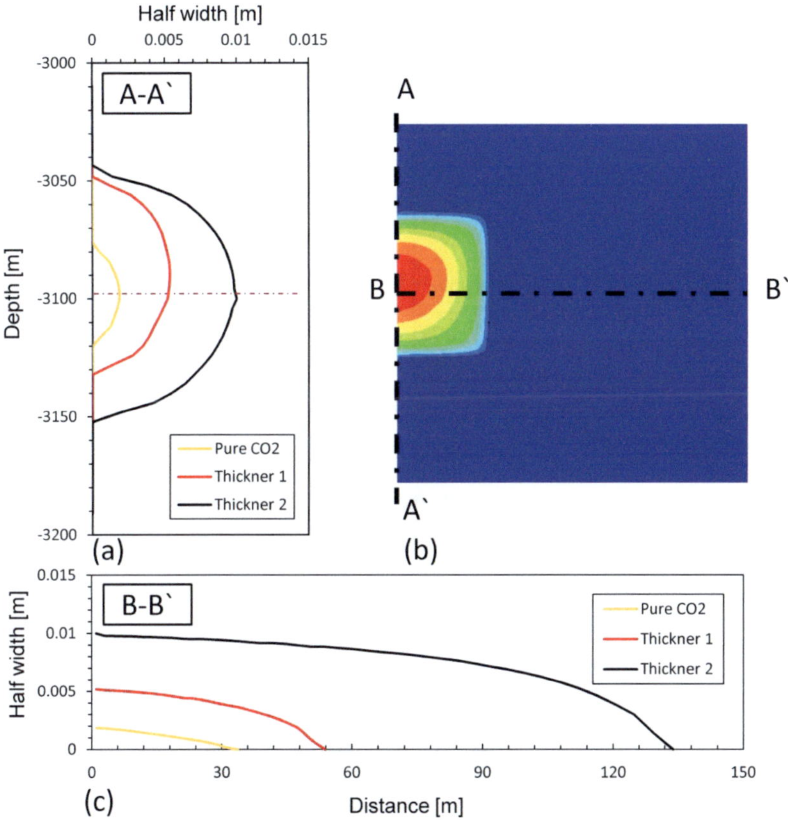

Figure 4.22. (a) Fracture profile created by different fluids along the dotted line A-A`, (B) fracture half width contour at 80min by Thickener 1 injection, (c) fracture profile created by different fluid along the dotted line B-B`

In the period of fracturing, the injection pressures by pure CO$_2$, Thickener 1 and 2 are recorded and plotted in Fig. 4.23a. The sawtooth nature of the injection pressure by pure CO$_2$ injection is because of the discrete element and coarse mesh. Otherwise, the main characteristic that injection pressure decreases from pure CO$_2$, Thickener 1 to Thickener 2 is clearly captured from these pressure cures. Because the fracture created by pure CO$_2$ is very small, there is not enough space for stress redistribution. As a consequent, a high pressure needs to overcome the compression stress on fracture wall. In order to catch the break down pressure, a time step low to 0.02s is used in the first ten seconds. The final results are compared in Fig. 4.23b. It is clear that the break down pressure shows a positive ratio with fracturing fluid viscosity, which is in accordance with the results observed by Zhang et al. (2017). The break down pressure reaches respectively to 75.3 MPa, 80.1 MPa and 84.0 MPa by pure CO$_2$, Thickener 1 and 2 injections.

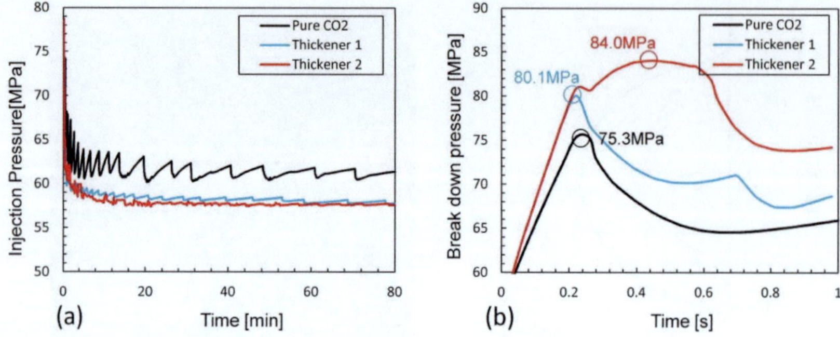

Figure 4.23. (a) Bottom hole pressure versus injection time, (b) break down pressure under different injection fluid

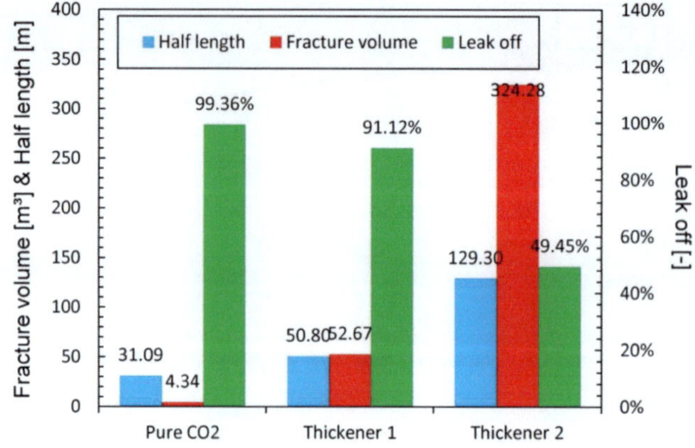

Figure 4.24. Comparing the simulated fracture half length, fracture volume and leak off under different fracturing fluid

The fracture half length, fracture volume as well as leak-off are compared in Fig. 4.24. Similar as the fracture half length, fracture volume increases from pure CO$_2$, Thickener 1 to Thickener 2 injection. The maximum fracture volume of 324.28m³ is created by Thickener 2 injection. Leakage is converted from the ratio between the stored and totally injected fluid mass. By pure CO$_2$ and Thickener 1 injection, most of the fracturing fluid is leaked. Even for Thickener 2 injection, still about 50% of injected content is leaked. But the fracture volume of 324.28 m³ exceeds the half volume of injected thickened CO_2, as a consequent of that the injected cold CO_2 is expanded in the fracture with relatively high temperature.

4.7. Comparing performance of thickened CO_2 with traditional fracturing fluid

As discussed above, Thickener 2 owning high performance in fracture creation is selected to compare with traditional fracturing fluid, including pure water and guar gum. The guar gum is injected at the temperature of 40°C with apparent viscosity about 100×10^{-3} Pa·s. Then fracture half width contours and pore pressure of host reservoir formation is shown in Fig. 4.25. The Thickener 2 could create a longer and wider fracture than pure water, but still shorter and narrower than the one created by guar gum, as guar gum has a much higher apparent viscosity. Maximum half length and half width due to guar gum injection could reach about 143m and 1.2cm, respectively. The pore pressure distribution by pure water injection is similar as the one by Thickener 1 injection in Fig. 4.18, because they have a comparable viscosity. While, the pore pressure by guar gum injection increases slightly in comparison with initial pressure (around 31MPa at injection point), as the leak-off rate is limited by high viscosity.

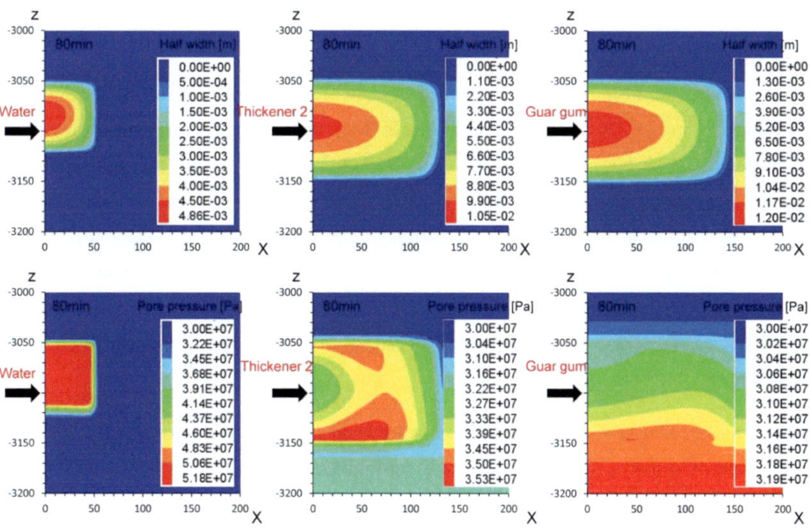

Figure 4.25. The results of the fracture half with and pore pressure with different fracturing fluid

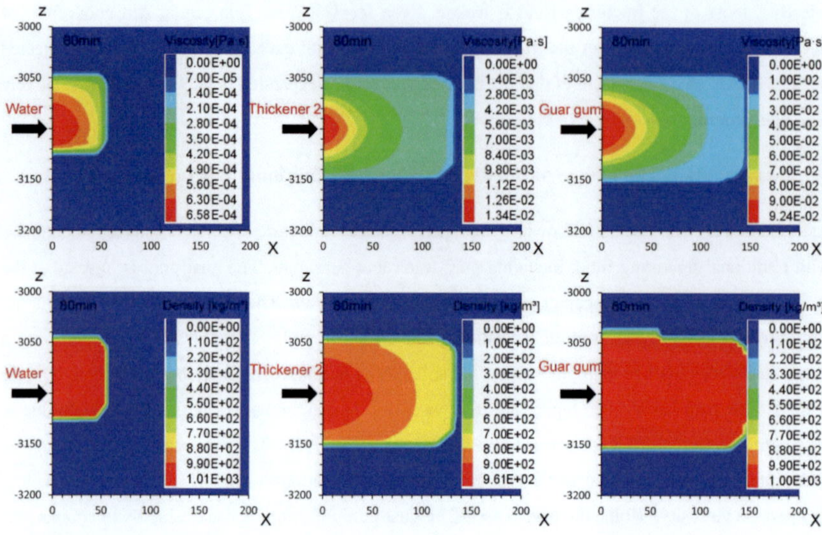

Figure 4.26. The results of the viscosity and fluid density with different fracturing fluid after fracturing

The viscosity of the fracturing fluid in fracture is exhibited in Fig 4.26, in which a comparable distribution is observed, decreasing with the distance away from injection point. But the maximum value is different, with the value of 0.65×10^{-3} Pa·s, 13.4×10^{-3} Pa·s, and 92.4×10^{-3} Pa·s from pure water, Thickener 2 and guar gum. At the end of fracturing, the density contours of fracturing fluid in fracture are also shown in Fig. 4.26. The density approximates $1000 kg/m^3$ and distributes uniformly in whole fracture, for the reason that water as well as guar water solution is not very compressible. In contrast, the density of CO_2 ranges from $961 kg/m^3$ to about $700 kg/m^3$, with the value decreasing form injection point to fracture boundary, because the density changes under influence of variable pressure and temperature.

The temperature contours due to pure water, Thickener 2 and guar gum are compared in Fig. 4.27. The lowest temperature of 40°C is found at injection point and equal to the injection temperature. All three have a similar distribution shape and the temperature increase from injection point to fracture tip. The difference is the propagated length of temperature drops area. The shortest propagated length of temperature drops area is due to pure water injection under the limit of fracture. In comparison with Thickener 2, the propagated length due to guar gum injection is much longer. There are two reasons for that: one is the fracture length created by guar gum is longer; another one is the injected cold guar gum has relatively high specific heat capacity, in another words, it takes more energy to heat the guar gum to the same temperature in comparison with CO_2.

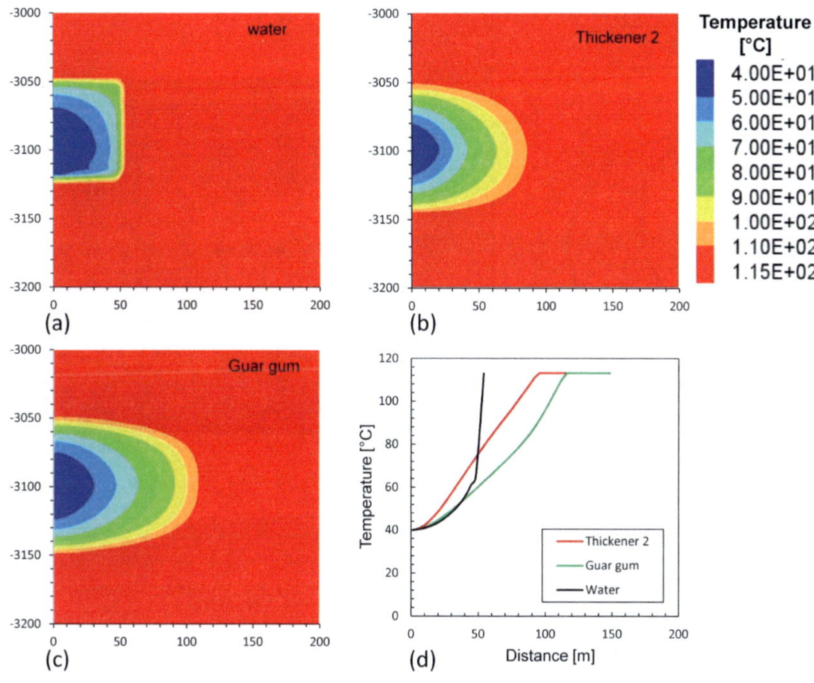

Figure 4.27. The temperature contours by (a) Water injection, (b) Thickener 2 injection, (c) Guar gum injection, (d) Temperature along a horizontal straight line crossing injection point

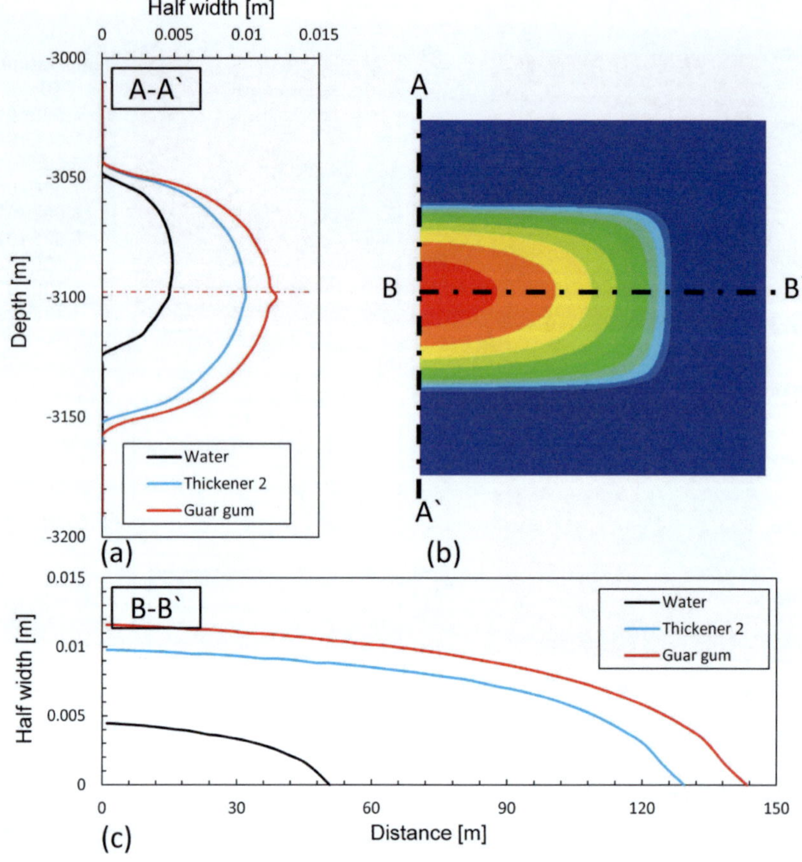

Figure 4.28. (a) Fracture profile created by different fluids along the dotted line A-A`, (B) fracture half width contour at 80min by Guar gum injection, (c) fracture profile created by different fluid along the dotted line B-B`

As shown in Fig. 4.28, the fracture width profiles created by different fluids injection are plotted along dotted line A-A` and B-B` respectively. The fracture profile created by pure water injection is apparently lower than other two fractures. The fracture width profiles due to Thickener 2 injection is very close to, but still slightly low as the one created by guar gum. It presents that the Thickener 2, namely the CO_2 thickened by surfactant, has a similar performance in fracture creation, comparing with traditional fracturing fluid guar gum.

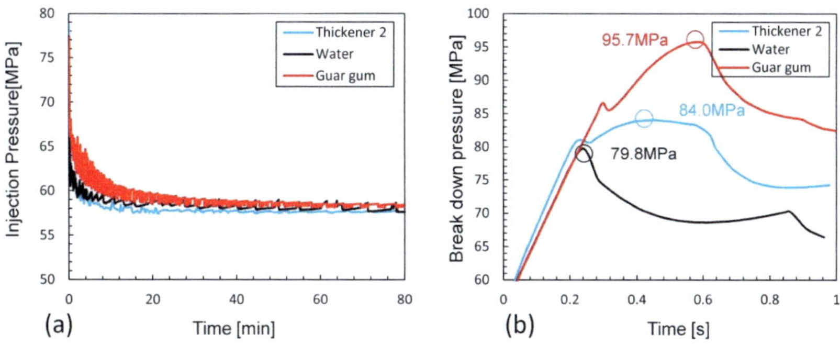

Figure 4.29. (a) Bottom hole pressure versus injection time, (b) break down pressure under different injection fluid

Fig. 4.29a shows the recorded injection pressure against time by different fluids injection, in which the injection pressure due to guar gum is beyond other two, with a dramatic fluctuation, because the guar gum with high viscosity needs overcome more resistance. This could be confirmed by the wider fracture half width in Fig. 4.25. Due to poor compressibility, the injection pressure fluctuates, when fracture tip propagates in a discrete element. However, the injection pressure by pure water, which owns the lowest viscosity, has a relatively high injection pressure in comparison with Thickener 2, since the relatively small fracture is not efficient for stress redistribution. Likewise, the break down pressure by different fluid injections is measured by a time step of 0.02s. A linear correlation between break down pressure and fracturing fluid viscosity is confirmed again. The break down pressure reaches to 79.8 MPa, 84.0 MPa and 95.7 MPa by pure water, Thickener 2 and guar gum injections, respectively.

The fracture half width is recorded at the injection point and plotted in Fig. 4.30. During injection of guar gum, the fracture width expands much quickly in first few minutes, because of high pressure at fracture initiation. Then it increases with a gradually slow down growing rate. A very similar tendency is found in the fracture width change by Thickener 2 injection. While, the fracture width due to pure water injection is almost stopped growing after fracture initiation. But the pressure fluctuation due to discrete element (see in Fig. 4.29a) is exactly reprinted on the fracture width change in Fig. 4.30.

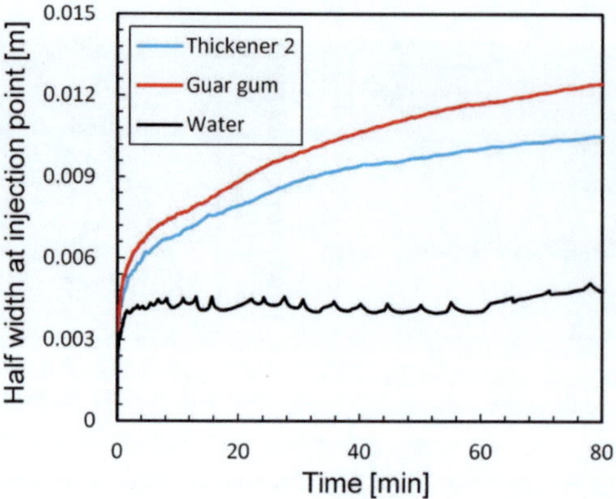

Figure 4.30. Temporal evolution of fracture half with under different injection fluid

The temporal variation of fracture volumes by pure water, Thickener 2 and guar gum are compared with injected volume in Fig. 4.31. Obviously, the final fracture volume decreases with fracturing fluid from guar gum, Thickener 2 to pure water. On the other hand, the leak-off rate, i.e. the difference between injection and fracture volume, rises with injection time due to an increasing leaked area (fracture area). For more information, Fig. 4.31a is partly magnified in to Fig. 4.31b. The fracture volume by Thickener 2 injection is sizably higher than the one by guar gum, even higher than injection volume, which is caused by the combined effect of significant volume expansion and almost negligible leakage at beginning. After that, the fracture volume increase gradually below fracture volume, because of increasing leakage.

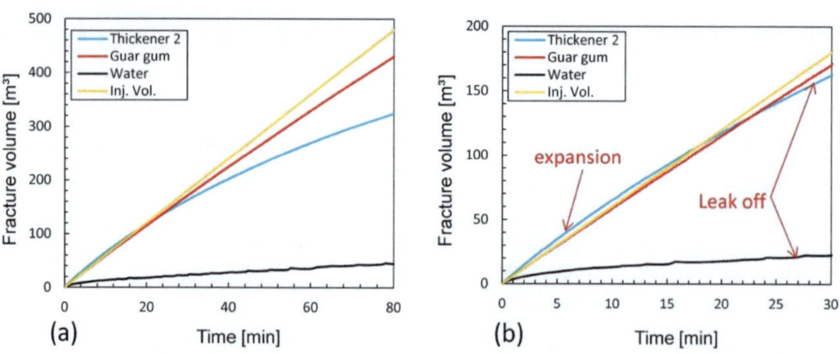

Figure 4.31. (a) Fracture volume versus injection time in the cases under different injection fluid, (b) partial magnification of the left figure

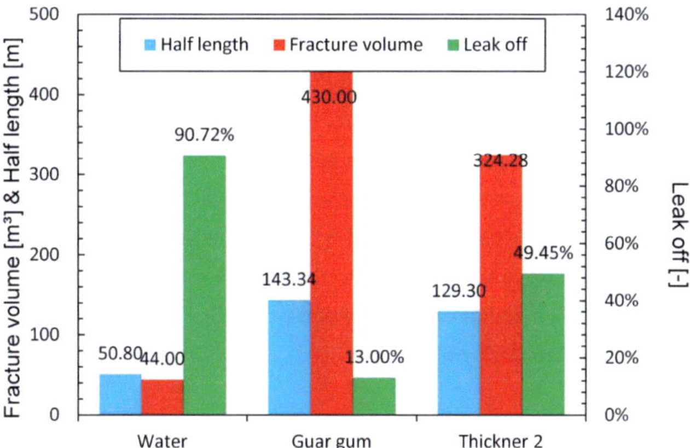

Figure 4.32. Comparing the simulated fracture half length, fracture volume and leak off under different fracturing fluid

Fig. 4.32 shows the fracture half length, fracture volume as well as leak-off by pure water, Thickener 2 and guar gum injection. The fracture half length and fracture volume have a similar tendency increase from pure water, Thickener 2 and guar gum, which is harmonized with the viscosity order of fracturing fluid. The maximum fracture length of 143m is created by guar gum injection, which is only about 10% longer than the one created by Thickener 2. If considering the fracture volume, the ratio between guar gum and Thickener 2 exceeds 32%, because of a much wider fracture created by guar gum. When the CO_2 is assumed incompressible, the fracture volume should be 240m³ calculated from the leak-off 49.5% and total injection volume of 480m³. Comparing with the real fracture volume of 324 m³, the fracture volume created by CO_2 expansion is accounted about 35%. Thus, the performance thickened CO_2 in fracture creation is comparable as guar gum in this model.

4.8. Thermal impact on fracture-making ability of thickened CO_2

As the property, especially the density and viscosity is closely correlated to fluid temperature, the performances of Thickener 2 under different thermal conditions are compared in this section. To characterize the thermal impact, several cases, Thickener 2 (thickened by surfactant) is injected at the temperature of 50°C under a non-isothermal condition (non-iso-50), Thickener 2 is injected into the isothermal reservoir with initial temperature of 40°C and 113°C (iso-40 & iso-113) respectively, are carried out and compared with the base case, which Thickener 2 (thickened by surfactant) is injected at the temperature of 40°C under a non-isothermal condition (non-iso-40). Two critical temperatures,

40 °C and 113°C are the injected temperature and temperature at reference depth of -3100m, respectively. Along the horizontal line crossing injection point, the temperature profiles as well as viscosity profiles after fracturing are exhibited in Fig. 4.33a and 4.33b, respectively. In the case of non-iso-40 and non-iso-50, the temperature profiles have a very similar tendency increases gradually with the distance away from injection point, but different start points. However, the temperature is hold constantly in whole fracture under isothermal condition, displaying as a horizontal line. The viscosity profiles show an exactly inverse tendency in comparison with temperature. A constant viscosity is found both in two isothermal cases. But the viscosity in case with temperature of 113°C is obviously higher than the one of 40°C. In non-isothermal cases, the viscosity decreases gradually from two values to the level determined by 113°C.

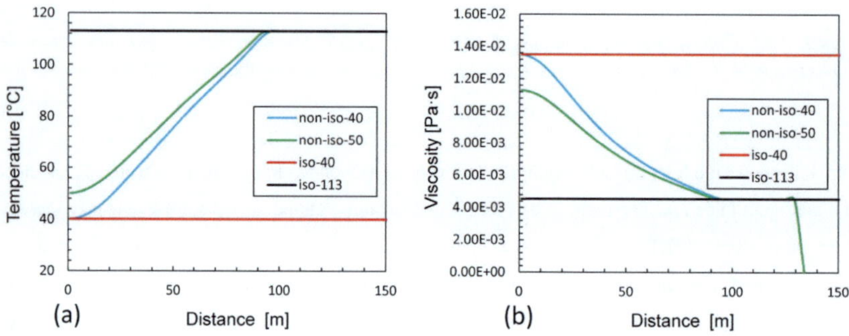

Figure 4.33. (a) Temperature (b) viscosity along the horizontal straight-line crossing injection point

On the same horizontal line, the fracturing fluid density in different cases is plotted in Fig. 4.34. The density profiles have a very similar tendency, comparing with viscosity in Fig. 4.32, except for the density of non-iso-40 near injection point. Even under same temperature, the density of non-iso-40 is higher than the one of iso-40, because the density of CO$_2$ is still associated with fracture pressure. A relatively high pressure is necessary to overcome high resistance caused by high viscosity in case of non-iso-40. The lowest density is found in case of iso-113, showing that the expansion effect is most significant. Under non-isothermal condition, the expansion effect is increased from injection point to fracture tip, in consistent with temperature in fracture.

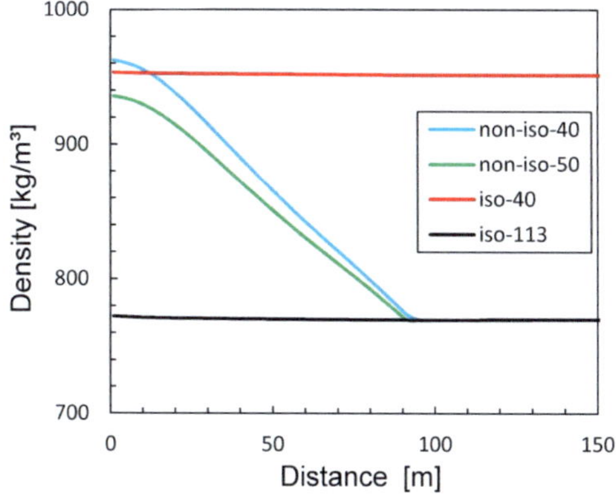

Figure 4.34. Density along the horizontal straight-line crossing injection point

The fracture half width profiles along the horizontal line crossing injection point are demonstrated in Fig. 4.35 in different cases. Under isothermal condition, the fracture in case of iso-40 is bigger than in case of iso-113, indicating that leak-off rate is the dominant factor to create a fracture in this fictitious model. The fracture width profiles in non-isothermal cases are not very different and lie between the fracture profiles of isothermal cases. However, the fracture widths of non-isothermal cases are still lower than the one of iso-113 near injection point, where the expansion effect of iso-113 is major in comparison with non-isothermal cases in this temperature drop area. Therefore, the spatial distribution of expansion effect and leak off rate in non-isothermal cases is uneven in fracture. Relatively high leak-off rate as well as expansion effect concentrates in the area close to fracture tip. And the fracture profiles are the competing results of leak-off and expansion.

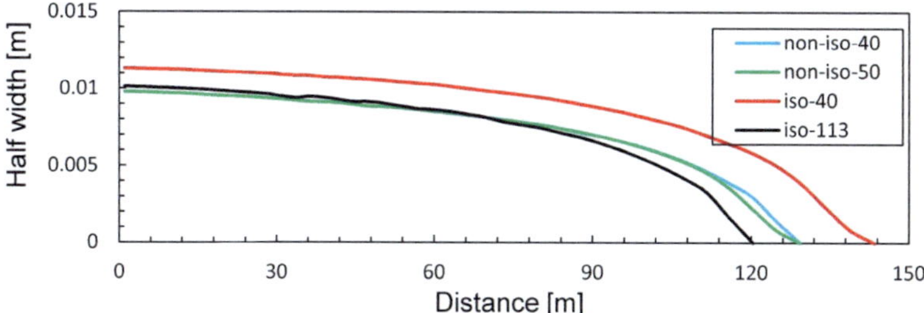

Figure 4.35. Fracture profile under different thermal conditions

The temporal fracture half width as well as fracture volume is shown in Fig. 4.36. In this discrete model, the maximum fracture half width increases in stepwise. The fracture length increases rapidly in the first few minutes, afterwards gradually slow down. The final fracture length decreases from iso-40, non-iso-40 & 50 to iso-113, which has been observed in Fig. 4.35. Regarding to the temporal fracture volume, it can be divided into two segments on the base of 20min. In the first segment, the order of fracture volume, iso-113>non-iso-40 & 50>iso-40, is observed, while it is exact opposite in the next segment. It shows that the temporal distribution of expansion effect and leak-off rate is uneven as well. At the beginning, the expansion effect is dominant in fracture creation, but with fracture propagation, leak-off rate will occupy the dominant place. Comparing the expansion effect, the leak-off, i.e. difference between injection and fracture volume, plays a particularly important role in fracture creation, especially in the second segment.

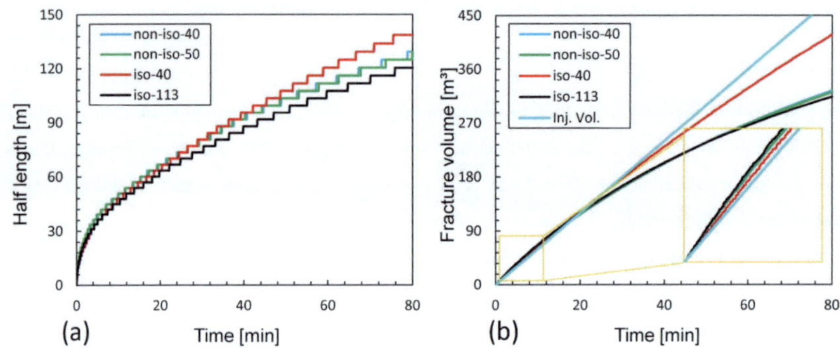

Figure 4.36. (a) Maximum half length (b) fracture volume under different thermal conditions

4.9. Optimization proppant densities and placement

Due to lower viscosity and density, the proppant-carrying ability of pure CO$_2$ as well as thickened CO2 is poor in comparison with traditional guar gum. To address the proppant-carrying issue, the simulation of different combinations (Thickener 2+2700kg/m³, Thickener 2+1750kg/m³ and Thickener 2+1250kg/m³) are simulated by using the newly developed model, and compared with the guar gum combining proppant density of 2700kg/m³. The injection rate as well as proppant concentration in fracturing fluid are shown in Fig. 4.37. In injected fracturing fluid, propant concentration increases in stepwise firstly after 30 min. Maximum injected volume concentration reaches about 19% after 70min and rough 37 m³ proppant is injected in every case.

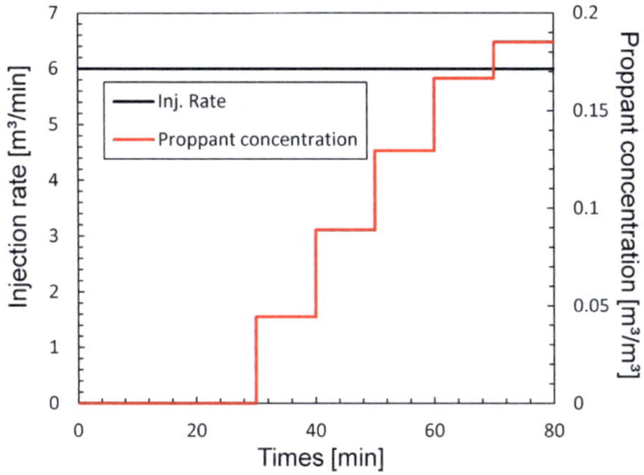

Figure 4.37. Injection plan and proppant volume concentration

The temporal fracture volumes of different cases are plotted in Fig. 4.38. After injection 480m³ fracturing fluid, all tree fracture volumes due to Thickener 2 reach all around 320m³, while guar gum creates a higher fracture volume about 430m³, because of lower leak-off. Then fracture well is shut in for fracture closure. Apparently, in all combinations of Thickener 2 injection, the fracture volumes decrease more quickly than guar gum. Only about 1 hour later, the fracture created by Thickener 2 is already closed. But it takes at least 5 hours for fracture closure in the case of guar gum, for the reason that Thickener 2 has a much significant leak-off rate driven by lower viscosity. After fracture closure, the final fracture volumes in all cases drop to the same level about 57m³, which is the minimum fracture volume can be supported by injected proppant volume (about 37 m³), accounted from the maximum proppant concentration at 0.65. It also confirmed that all fractures are fully closed. In addition to this, the temporal fracture volume curves due to Thickener 2 injection overlap each other in whole process, indicating that the fracture propagation and closure is not influenced by proppant density, but mainly determined by the property of fracturing fluid.

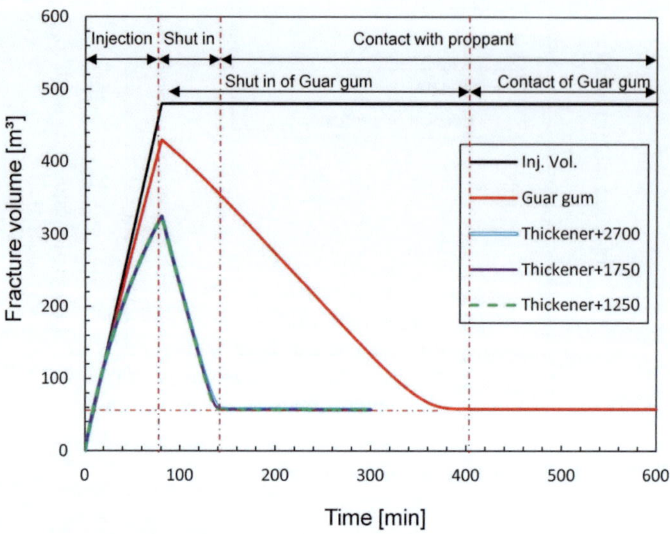

Figure 4.38. Temporal evaluation of fracture volume from injection to fracture closure

The temporal fracture volume variation could be explained fundamentally by the temperature and viscosity change during fracture closure. Along the horizontal line crossing injection point, the temperature and viscosity profiles of fracturing fluid (Thickener 2) are drawn in Fig. 4.39 respectively, at different time points during fracture closure. Because no more cold fracturing fluid is injected after well shut in, the temperature of fracturing fluid in fracture recovers gradually through heat conduction from host rock. As shown in Fig. 4.39a, the temperature recovers with a gradually slow rate. About 100min after fracturing, the temperature at injection point is already recovered from 40°C to 105°C, but the final temperature is still slightly lower than initial temperature, because of energy balance. It should be pointed out that the temperature drop area propagates continuously even after fracturing, due to unevenly spatial leak-off rate, the cold fracturing fluid will be further transferred to fracture tip to fill the leaked content. Correspondingly, the viscosity (in Fig. 3.39b) shows an inverse tendency, decreasing gradually with time as well as distance away from injection point. Only 100min late, the viscosity of Thickener 2 is almost recovered.

As shown in Fig. 4.40, the temperature and viscosity profiles due to Thickener 2 injection are compared with guar gum at different time points. In the case of guar gum injection, the temperature recovery is similar, but much slower in comparison with the one by Thickener 2 injection, one of probable reason is that thickened CO$_2$ owns a low specific heat capacity in comparison with guar water solution. In another word, the temperature recovery of Thickener 2 is quicker than guar gum under the same heat conductivity. It will accelerate the gel breaking of Thickener 2, which is evidenced also by the viscosity profiles in Fig. 4.40b. About 100min after fracturing, the temperature of Thickener 2 injection is almost recovered, while the one of guar gum injection is still high. In spite of this, the viscosity of guar gum at original reservoir temperature is apparently higher than Thickener 2.

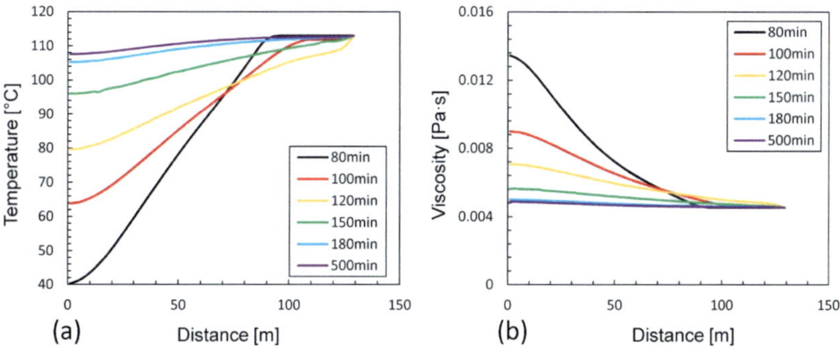

Figure 4.39. (a) Temperature and (b) viscosity along the horizontal straight line crossing injection point in the cases with Thickener 2+proppant 2700kg/m³ injection

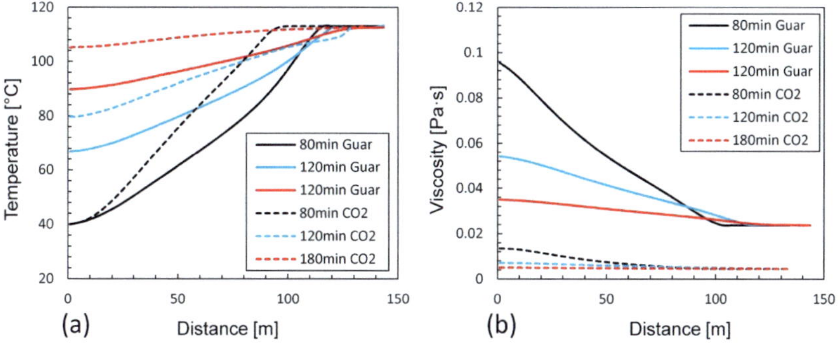

Figure 4.40. Comparing (a) temperature and (b) viscosity along the horizontal straight line crossing injection point in the cases with Thicker 2 and Guar gum injection respectively

Before and after shut in, the vertical fracture width profiles of different cases along A-A` are compared in Fig. 4.41, respectively. Obviously, the light proppant density could improve proppant placement. Form the fracture width profile, the supported fracture width by light proppant is higher than the other two cases, because the settling velocity of proppant in fracture is dependent on proppant density. Heavy proppant falls much more rapidly and concentrates at the bottom fracture, thus the supported fracture width by Thickener 2+2700kg/m^3 is shorter and wider. The proppant placement is even better than guar gum, because in guar gum injection, the proppant has enough time and space to settle down in a relatively wide fracture, besides the guar gum owns a slow gel breaking speed.

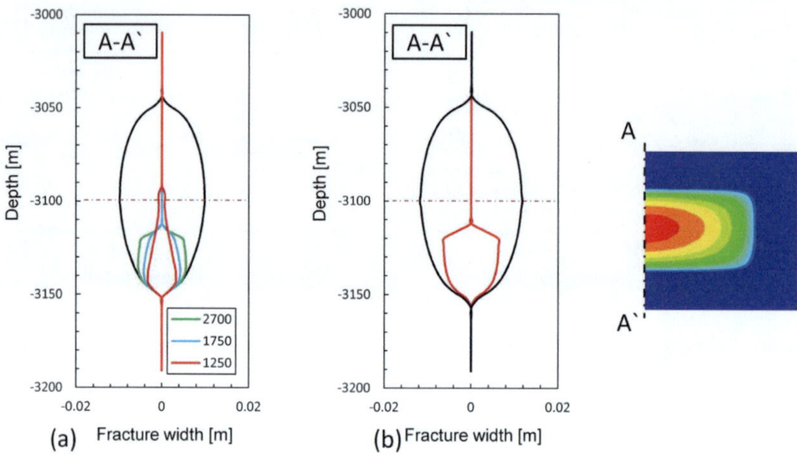

Figure 4.41. Comparing fracture profile along dotted line A-A` after fracturing and after closure, (a) Thickener with different proppant densities, (b) Guar gum with proppant density of 2700kg/m^3

After fracture closure, the proppant placement contours of different cases are illustrated in Fig. 4.42. Comparing different proppant transported by Thickener 2, light proppant can be transported in deeper fracture than heavy proppant, as the light proppant with lower settling velocity has enough time to be transported into fracture. On the hand, light proppant will support a high fracture, as shown in Fig. 4.41a as well. Comparing with proppant transported by guar gum, the proppant is placed in the bottom fracture much closer to injection point side, as a wider fracture created by guar gum injection provide more time and space for proppant settling down, which is confirmed by final supported fracture. The maximum supported fracture half width about 0.63cm is found in the cases of guar gum injection. Additionally, only in the case of Thickener 2+1250kg/m^3, the final supported fracture connects directly with injection point. Therefore, the combination of Thickener 2+1250kg/m³ is recommended for fracturing in a comparable tight gas reservoir.

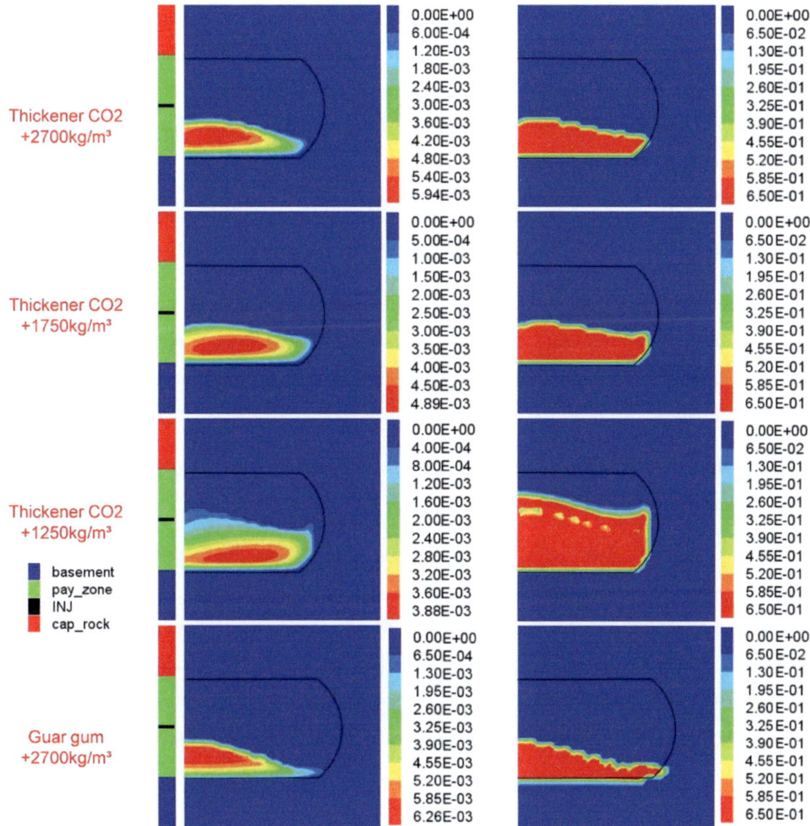

Figure 4.42. Comparing fracture half with contours and proppant concentration contours after closure under different combination of fracturing fluid and proppant density

4.10. Summary

In this chapter, a pre-defined potential fracture plane, which is perpendicular to the minimum stress, is inserted in the middle of host reservoir elements. On this potential fracture plane, numerical models, including fracture model, proppant transport model and multi-flow model, are developed and integrated in the popular THM framework TOUGH2MP-FLAC3D, to study performance of an alternative fracturing fluid, supercritical CO$_2$, in thigh gas reservoir. These implemented models have been verified separately. Besides, the property (mainly temperature-dependent viscosity) of different fracturing fluids is defined, like pure CO_2, different thickened CO_2s and guar gum. Then, the fracturing by using different fracturing fluids has been carried out on a fictitious model with the properties of typical tight gas reservoir. Their performances in fracture creation are compared based on the simulated results. As the property of CO_2 is closely correlated to temperature, the thermal effect has been discussed under different thermal conditions. Additionally, proppant with different densities is used to improve the weak proppant-carrying ability of CO_2. Some important conclusion is summarized in following

Pure CO_2 is not efficient to create a fracture in such tight gas reservoir. However, the fracture making ability of CO_2 can be improved significantly, by adding CO_2 thickener, like surfactant.

In comparison with guar gum, the thickened CO_2 has higher leak-off rates. But if considering the expansion effect of CO_2 in fracture, thickened CO_2 could show a comparable performance in fracture creation as traditional guar gum. In this work, fracture-making ability could be improved about 37% by CO_2 expansion.

The break down pressure shows linear correlation with fracturing fluid viscosity. The break down pressure rises with increasing fracture fluid viscosity.

The fracture shape is determined by the combination of CO_2 expansion and leak-off rate. In non-isothermal condition, the temporal and spatial distribution of the leak-off rate and expansion effect is non-uniform. High leak-off rate and expansion effect is found around fracture tip, due to high temperature. At the beginning of fracturing, the expansion effect plays a critical role in fracture creation. With continuous injection, the dominant factor in fracture creation is gradually switched to leak-off. Overall, for the influence on fracture-making ability, leak-off is larger than expansion in whole process.

The thickened CO_2 with light proppant could achieve a better proppant placement than heavy proppant, even better as the one transport by guar gum. Due to low specific heat capacity, thickened CO_2 has a quick gel breaking speed, resulting that the fracture can be supported before proppant fully settling to fracture bottom.

5. Geothermal energy exploitation with supercritical CO$_2$

5.1. Basic idea

Before heat extraction in a geothermal reservoir, especially HDR reservoir, the permeability of virgin rock need be enhanced usually by hydraulic fracturing, to improve the heat production efficient. The basic mechanism of hydraulic fracturing can be explained in Fig. 5.1. The increasing pressure due to fluid injection can change the stress state, which further induce micro cracks namely damage on the host rock. Apparently, the micro cracks are directional. Its direction is closely correlated to the normal strain e.g. tensile strain on cracks wall. These directional cracks or damage can be used to predict the equivalent anisotropic permeability of this fractured rock. Lee and Ghassemi (2009) performed a fundamental numerical study on such stress-dependent permeability based on continuum damage mechanics, to study the fracture propagation in geothermal reservoir. In fact, stress-dependent permeability could be explained by the stress-dependent micro cracks (Massart and Selvadurai, 2014), in which permeability was evaluated by using the Stokes' flow between two parallel plates. Average method was applied to calculate permeability, based on a representative element volume (REV), whereas such micro model is inefficient to solve the large-scale problem. Conversely, a macro model that is developed by Souley et al. (2001) explains the relation between macroscopic damage tensor and induced permeability in granite. But only an isotropic tensorial character of permeability was taken into account in this model. Then Maleki and Pouya (2010) established an anisotropic damage-permeability model based on loading types in a brittle geomaterials.

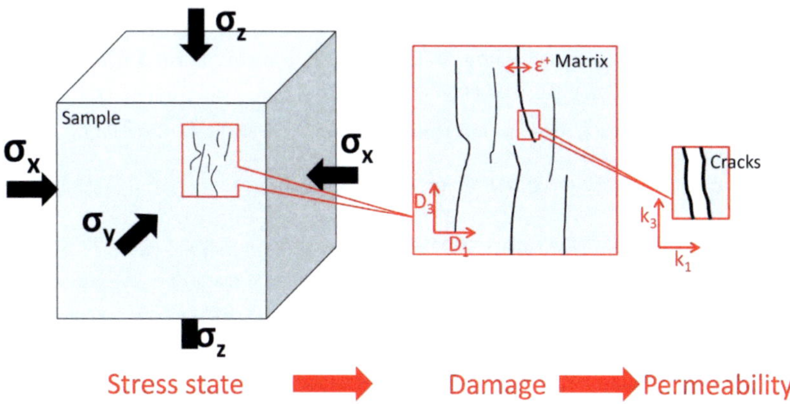

Figure 5.1. Correlation between stress state, damage and permeability

Combining this anisotropic damage-permeability model, a 3D continuum damage model has been developed by using FDM on the fundamental THM framework introduced in chapter 3, to study the permeability change during hydraulic fracturing in a brittle HDR reservoir. Comparing with other methods, this continuum approach performs more comprehensively and provides an ease for implementation in existing geomechanical codes (FLAC3D), computing speed and possibility to apply statistical knowledge directly instead of creating a representation for a fracture network. The damage model and anisotropic permeability have been verified through series triaxial test results separately. In addition to this, feasibility of the damage-permeability model in HDR hydraulic fracturing is verified also by a cyclic hydraulic fracturing test on hard rock. To gain more understanding of CO_2 fracturing in HDR reservoir, parametric study on influence factors has been conducted on a simple 2D model. The a field scale simulation has been carried out based on the geological from planned Dikili EGS for over 30 years, in which the driven pressure, produced temperature as well accumulated net thermal energy haven compared under different injection rates. Besides, the sequestrated CO_2 during lifetime of geothermal project has been evaluated.

5.2. Description of anisotropic damage-permeability model

The mathematic description of the damage model and permeability model has been introduced by previous works (Liao et al. 2019b; Liao et al. 2020b) and is briefly reviewed in this section. It is believed that the damage takes place in reservoir during hydraulic fracturing, mainly due to variable effective stress, which is given in following formula of pore pressure.

$$\sigma' = \sigma + \alpha P \boldsymbol{I} \tag{5.1}$$

Where σ' and σ are effective and total stress tensor respectively (compressive is negative) [Pa]. α is the biot coefficient [-]. P is the pore pressure in the pore geo-material [-]. $\boldsymbol{I}$ is an unite tensor. According to previous work by Chiarelli et al. (2000), the effective stress can be expressed by the combination of traditional Hooks' law with terms of damage ($\boldsymbol{D}$) in elastic damage model.

$$\sigma' = \lambda_0 tr\varepsilon^e \boldsymbol{I} + 2\mu_0\varepsilon^e + a_1(tr(\varepsilon^e \boldsymbol{D})\boldsymbol{I} + tr(\varepsilon^e)\boldsymbol{D}) + a_2(\varepsilon^e \boldsymbol{D} + \boldsymbol{D}\varepsilon^e) \tag{5.2}$$

Where λ_0 and μ_0 are the elastic Lame constant and shear modulus in an undamaged state respectively [Pa]. Two model parameters a_1 and a_2 describe the degradation of the elastic properties caused by induced damage. ε^e is an elastic tensor [-], which is the sum of mechanical term ε^e_M and thermal term ε^e_T [-].

$$\varepsilon^e = \varepsilon^e_M + \varepsilon^e_T \tag{5.3}$$

The thermal deformation is assumed isotropic. Referring to Cooper and Simmons (1997), the thermal strain increment can be written as

$$\varepsilon_T^e = -\beta \, \Delta T I \tag{5.4}$$

Here, β is the thermal expansion coefficient [1/°C]. ΔT is the temperature increment [°C]. For plastic deformation, the total strain involves elastic strain (ε^e) and plastic strain (ε^p) as following

$$\varepsilon = \varepsilon^e + \varepsilon^p \tag{5.5}$$

Substitute Eq. 5.5 into Eq. 5.2, the Eq. 5.2 can be simplified as

$$\sigma' = \mathbb{C}(D): \varepsilon^e = \mathbb{C}(D): (\varepsilon - \varepsilon^p) \tag{5.6}$$

Where $\mathbb{C}(D)$ is the effective elastic stiffness tensor for damage material, which is given by Halm and Dragon (1996) in the form of a fourth order symmetric tensor.

$$\mathbb{C}_{ijkl} = \lambda_0 \delta_{ij}\delta_{kl} + \mu(\delta_{ik}\delta_{jl} + \delta_{il}\delta_{jk}) + a_1(\delta_{ij}D_{kl} + D_{ij}\delta_{kl}) + \frac{a_2}{2}(\delta_{ik}D_{jl} + \delta_{il}D_{jk} + D_{ik}\delta_{jl} + D_{il}\delta_{jk})$$

Take the derivative on both sides of Eq. 5.6, the stress increment tensor can be transforms into

$$\Delta\sigma' = \mathbb{C}(D): (\Delta\varepsilon - \Delta\varepsilon^p) + \Delta\mathbb{C}(D): \varepsilon^e \tag{5.7}$$

Where $\Delta\varepsilon$ and $\Delta\varepsilon^p$ are the total and plastic strain increment tensor [-]. Increment $\Delta\mathbb{C}(D)$ is the derivative of effective elastic stiffness tensor with respect to the damage tensor (Eq. 5.8)) [Pa]. The specific sixth order increment tensor is given by Yang et al. (2013).

$$\Delta\mathbb{C}(D) = \frac{\partial \mathbb{C}(D)}{\partial D}: \Delta D \tag{5.8}$$

$$\Delta\mathbb{C}_{jlkl} = \frac{a_1}{2}\left[\delta_{jl}(\delta_{km}\delta_{ln} + \delta_{kn}\delta_{lm}) + (\delta_{in}\delta_{jm} + \delta_{im}\delta_{jn})\delta_{kl}\right] \times \Delta D_{mn} + \frac{a_2}{4}\left[\delta_{ik}(\delta_{jm}\delta_{ln} + \delta_{jn}\delta_{lm}) + \delta_{il}(\delta_{jm}\delta_{kn} + \delta_{jn}\delta_{km}) + (\delta_{im}\delta_{kn} + \delta_{in}\delta_{km})\delta_{jl} + (\delta_{im}\delta_{ln} + \delta_{in}\delta_{lm})\delta_{jk}\right] \times \Delta D_{mn}$$

Wherein, the plastic strain increment is defined by the plastic flow rule in following

$$\Delta\varepsilon^p = \lambda_p \frac{\partial g}{\partial \sigma} \tag{5.9}$$

Here λ_p is the plastic multiple. g is the potential function. The failure criterion (f) defines the boundary between elastic and plastic state. To judge whether the plastic deformation takes place, the failure criterion will be checked under each stress state. In this thesis, the popular constitutive model-ubiquitous-joint model is adopted to describe the jointed hard rock in geothermal reservoir. As shown in Fig. 5.2, the virgin rock contains many joint structures with dominant stochastic direction in **n**. Two classic failure criterions, Mohr criterion and tensile criterion, are considered on the matrix as well as joints respectively. All the failure criterions with corresponding potential functions are summarized in Eq. 5.10, in which the subscript m and j represent matrix and joint, respectively. The Mohr and tensile failure are denoted by *s* and *t* separately. In particular, both failure criterion of joint are generated in a local coordination on joint area, which they of matrix are generated in global coordination.

$$\begin{cases} F_m^s = \sigma_1' - \sigma_3' N_{\varphi_m} \eta(k_m) + 2c_m \sqrt{N_{\varphi_m}} \eta(k_m) \\ F_m^t = \sigma_3' - \sigma^t \eta(k_m) \\ F_j^s = \tau - \sigma'_{3'3'} \tan\varphi_j \eta(k_m) + c_j \eta(k_m) \\ F_j^t = \sigma'_{3'3'} - \sigma_j^t \eta(k_m) \\ G_m^s = \sigma_1' - \sigma_3' N_{\theta_m} \\ G_m^t = \sigma_3' \\ G_j^s = \tau + \sigma'_{3'3'} \tan\theta_j \\ G_j^t = \sigma'_{3'3'} \end{cases} \tag{5.10}$$

Where σ_1' and σ_3' are the principal effective stress ($\sigma_1' < \sigma_3'$, compressive stress is negative) [Pa]. σ^t is the tensile strength [Pa]. φ and θ are the friction angle and dilation angle respectively [°]. c is the cohesion [Pa]. τ is shear stress acting on the joint plane [Pa]. $\sigma'_{3'3'}$ is the effective normal stress on the joint plane [Pa]. The term N_{φ_m} is defined as $N_{\varphi_m} = \frac{1+\sin\varphi_m}{1-\sin\varphi_m}$. Function $\eta(k_m)$ is used to describe the hardening-softening behavior during loading.

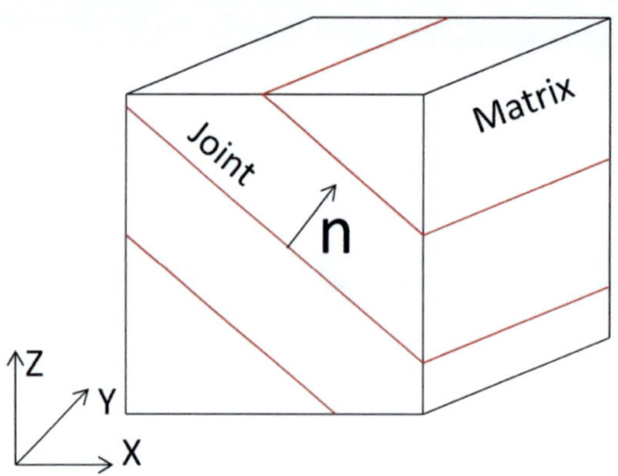

*Figure 5.2. Illustration of jointed matrix with joints dominant in **n** direction (Liao et al. 2020b)*

According to the work of Chen et al. (2015), the brittle rock shows a hardening behavior that relating to the plastic strain, a softening behavior to that relating to the damage. Based on the results, a function $\eta(k_m)$ in following is introduced to express hardening/softening behavior.

$$\eta(k_m) = (1 - tr\mathbf{D})^R \left[1 + (\eta_m - 1)\frac{k_m}{b_1 + k_m}\right] \tag{5.11}$$

Where R is a ratio to control the softening rate [-]. b_1 is a hardening parameter [-] and η_m is defined as the ratio between the final and initial plastic surface. The strain-hardening parameter k_m is determined on the base of three principal plastic strains ($\varepsilon_1^p, \varepsilon_2^p, \varepsilon_3^p$) in the following equation.

$$k_m = \frac{1}{3}\sqrt{\left(\varepsilon_1^p\right)^2 + \left(\varepsilon_2^p\right)^2 + \left(\varepsilon_3^p\right)^2} \tag{5.12}$$

Chiarelli et al. (2003) presented a simple damage criterion, based on the generalized tensile strain $\boldsymbol{\varepsilon}^+$, which is the positive cone of the total strain tensor. In this thesis, the damage criterion is modified as following by introducing an upper limit of damage (D_{max}).

$$f(\boldsymbol{\varepsilon}^+, \mathbf{D}) = \sqrt{\boldsymbol{\varepsilon}^+ : \boldsymbol{\varepsilon}^+} - \left[r_0 + \frac{r_1}{(1 - \frac{tr\mathbf{D}}{D_{max}})^{k_d}} tr\mathbf{D}\right] \tag{5.13}$$

Where two parameters r_0 and r_1 define the initial damage threshold and damage evolution rate [-] respectively. The value of upper limit D_{max} can be larger than one, as it compares with the trace of damage tensor. The constant k_d controls the damage evolution rate, which slows down with damage increasing to the upper limit. The expression of damage increment can be written in Eq. 5.14. Set the derivative of Eq. 5.13 equal to zero, the damage multiple can be derived in Eq. 5.15.

$$\Delta\mathbf{D} = \lambda_d \frac{\boldsymbol{\varepsilon}^+}{\sqrt{\boldsymbol{\varepsilon}^+ : \boldsymbol{\varepsilon}^+}} \tag{5.14}$$

$$\lambda_d = \frac{(D_{max} - tr\mathbf{D})^{k_d+1}}{D_{max}^{k_d+1} + D_{max}^{k_d} tr\mathbf{D}(k_d - 1)} \cdot \frac{\boldsymbol{\varepsilon}^+ : \Delta\boldsymbol{\varepsilon}^+}{r_1 \cdot tr\boldsymbol{\varepsilon}^+} \tag{5.15}$$

The permeability change during hydraulic fracturing can be divided into two parts. One is caused by effective stress, another one by damage. Considering the permeability change under a varying effective stress, the pore, or preexisting cracks even induced cracks can be compressed under a increasing effective stress, showing in declined permeability, and vice versa. As damage in this thesis is converted from the cracks characteristic-tensile strain, the permeability due to damage can be approximated by damage cubic law, which is widely used to characterize flow behavior between fracture apertures. Here, the permeability function suggested by Zhang (2016) is modified in the term of effective means stress and cubic damage in following

$$K = K_0[c_1 + (1 - c_1)exp(-\gamma\langle\sigma_m'\rangle)] + K_0 c_2 (tr\mathbf{D})^3 \tag{5.16}$$

Where K_0 is the original permeability of virgin rock at zero stress [m²]. c_1 is the ratio of the minimum and original permeability [-]. γ is a parameter characterizing the dilatability of interconnected cracks [MPa⁻¹]. c_2 is ratio of the maximum damage-induced permeability and original permeability [-]. The effective stress induced permeability (K_{es}) and damage induced permeability (K_D) can be written in following

$$K_{es} = K_0[c_1 + (1 - c_1)exp(-\gamma\langle\sigma'_m\rangle)] \tag{5.17}$$

$$K_D = K_0c_2(tr\mathbf{D})^3 \tag{5.18}$$

In this thesis, the effective stress induced permeability is treated as isotropic, while the damage induced permeability is anisotropic, whose direction is dependent on the induced cracks. Further, cracks' direction is tightly correlated to different loading types. As illustrated in Fig. 5.3, Maleki and Pouya (2010) used ellipsoid to exhibit the normalized permeability in vertical direction under different loading types, in which the induced permeability was evaluated based on cracks density in a radial rotating cut plane. Entirely 36 cut planes, each with 10° rotation angle, were evaluated. Radial axis is the normalized permeability that equals to the ratio of obtained permeability to the maximum permeability. The normalized permeability distributes itself symmetrically with the cut plane having 0°, which corresponds to axial direction of sample cylinder. In compression case (Fig. 5.3(a)), most of the cracks lie in vertical direction. Therefore, maximum induced permeability is measured in vertical direction. Yet, in Fig. 5.3(b), most of cracks are measured in horizontal direction under extension case. The permeability in case of isotropic loading is assumed isotropic. Through a simple linear interpolation method, the permeability direction of a general triaxial case can be interpreted by a combination of different loading types. According to Maleki and Pouya's work (2010), the expression of anisotropic permeability tensor relating to induced damage can be written as

$$\mathbf{K}_D = \frac{K_D}{tr\mathbf{D}}\{[\alpha_e(d_1 - d_2) + 2\alpha_c(d_2 - d_3) + d_3]\delta + (d_1 - d_2)(1 - 3\alpha_e)\underline{e}_1\otimes\underline{e}_1$$
$$+ 2(d_2 - d_3)(1 - 3\alpha_c)\underline{e}_3\otimes\underline{e}_3\} \tag{5.19}$$

Where d_1, d_2, and d_3 are three principal values of $\mathbf{D}$ with $d_1 \geqslant d_2 \geqslant d_3$. e_1 e_2 and e_3 are directions corresponding to three principal damages, respectively. α_c and α_e are parameters from compression case and extension case respectively. $\alpha_c = \frac{1}{4}$ and $\alpha_e = \frac{5}{12}$ refers to (Maleki and Pouya, 2010).

Then, the total permeability tensor can be given in following

$$\mathbf{K} = K_{es}\mathbf{I} + \mathbf{K}_D \tag{5.20}$$

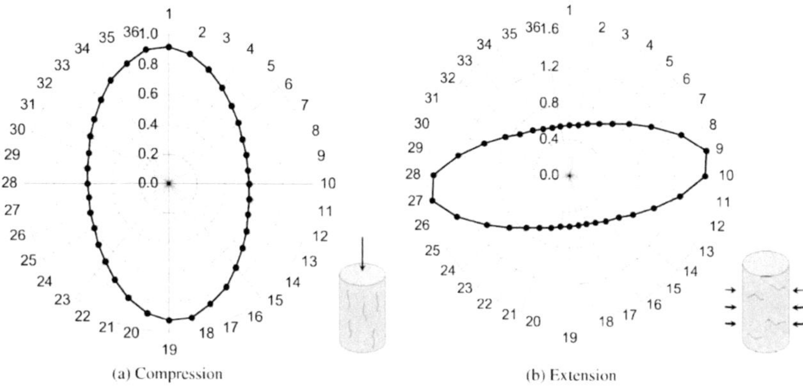

(a) Compression   (b) Extension

Figure 5.3. Normalized permeability ellipsoid in radial plane along vertical direction of different types (Maleki and Pouya, 2010)

5.3. Implementation of numerical models and working schema

Both the anisotropic damage model and permeability model were integrated in FLAC3D, which is a popular commercial code by using finite difference method (FDM). Similarly, the model was discretizated in space as well as in time. The numerical implementation of damage and permeability model was introduced on one element of discrete model. At the new time step (t+1), the total strain ($\varepsilon_{ij}(t+1)$) on element center is firstly converted from displacement on itself grid point, with the help of isoparameter method. The new strain increment ($\Delta\varepsilon_{ij}^e(t+1)$), which is recognized initially all as elastic strain increment, is calculated in Eq. 5.21. The elastic strain at previous step ($\varepsilon_{ij}^e(t)$) is obtained from difference of total strain and plastic strain from last time step.

$$\Delta\varepsilon_{ij}^e(t+1) = \varepsilon_{ij}(t+1) - \varepsilon_{ij}(t) \tag{5.21}$$

$$\varepsilon_{ij}^e(t) = \varepsilon_{ij}(t) - \varepsilon_{ij}^p(t) \tag{5.22}$$

With the current total strain ($\varepsilon_{ij}(t+1)$), the total generalized tensile strain ($\varepsilon^+(t+1)$) can be updated by

$$\boldsymbol{\varepsilon}^+(t+1) = \sum_{i=1}^3 \langle \varepsilon_i(t+1) \rangle V_i \otimes V_i \tag{5.23}$$

Here the $\varepsilon_i(t+1), V_i$ (i=1,2,3) are the principal strain component and its corresponding unite direction vector. $\langle\,\rangle$ is Macaulay brackets given in following

$$\langle \varepsilon_i(t+1) \rangle = \begin{cases} \varepsilon_i(t+1) & if\ \varepsilon_i(t+1) > 0 \\ 0 & if\ \varepsilon_i(t+1) \le 0 \end{cases}$$

Then the damage criterion (Eq. 5.13) is checked on the base of new total generalized tensile strain. The new generalized tensile strain increment ($\Delta\varepsilon_{ij}(t+1)$) need be solved according to Eq. 5.23, only when the damage criterion is satisfied. Otherwise, the new damage increment ($\Delta D_{ij}(t+1)$) is set as zero. The achieved generalized tensile strain as well as current increment is further used to evaluate current damage multiplication ($\lambda_d(t+1)$). Subsequently, the new damage increment is acquired by following equation

$$\Delta D_{ij}(t+1) = \lambda_d(t+1)\frac{\varepsilon_{ij}^+(t+1)}{\sqrt{\varepsilon^+(t+1):\varepsilon^+(t+1)}} \tag{5.24}$$

The new effective stress increment ($\Delta\sigma_{ij}'(t+1)$) is calculated in Eq. 5.25 with the help of new strain increment, old elastic strain, effective elastic stiffness tensor ($\mathbb{C}_{ijkl}(t)$) and its increment ($\Delta\mathbb{C}_{ijklmn}(t+1)$), in which the last two parameter are updated based on old damage ($D(t)$) and new damage increment ($\Delta D(t+1)$), respectively.

$$\Delta\sigma_{ij}'(t+1) = \sum_{k=1}^3\sum_{l=1}^3 \mathbb{C}_{ijkl}(t+1)\Delta\varepsilon_{ij}^e(t+1) + \sum_{k=1}^3\sum_{l=1}^3\sum_{m=1}^3\sum_{n=1}^3 \Delta\mathbb{C}_{ijklmn}(t+1)\varepsilon_{ij}^e(t)$$

$$\tag{5.25}$$

The elastic trial stress ($\sigma_{ij}'(t+1)^{trial}$) on element center is calculated by

$$\sigma_{ij}'(t+1)^{tial} = \sigma_{ij}'(t) + \Delta\sigma_{ij}'(t+1) \tag{5.26}$$

With this trial stress, the failure criterion (summarized in Eq. 5.10) need be checked on both matrix and joint. If any plastic failure occurs, the relevant plastic strain increment ($\Delta\varepsilon_{ij}^p(t+1)$) will be calculated according to Eq.5.9. The corresponding stress relating to the plastic strain increment need be removed and the new tress state ($\sigma_{ij}'(t+1)$) is expressed in following

$$\sigma_{ij}'(t+1) = \sigma_{ij}'(t+1)^{trail} - \sum_{k=1}^3\sum_{l=1}^3 \mathbb{C}_{ijkl}(t)\Delta\varepsilon_{ij}^p(t+1) \tag{5.27}$$

The new stress state with current pore pressure and temperature are assigned on the element, for the new strain in the next time step. Additionally, the plastic strain ($\varepsilon_{ij}^p(t+1)$) and damage ($D_{ij}(t+1)$) are accumulated according to Eq. 5.28 and Eq. 5.29 respectively.

$$\varepsilon_{ij}^p(t+1) = \varepsilon_{ij}^p(t) + \Delta\varepsilon_{ij}^p(t+1) \tag{5.28}$$

$$D_{ij}(t+1) = D_{ij}(t) + \Delta D_{ij}(t+1) \tag{5.29}$$

The principal damage component is obtained by

$$\tilde{D}(t+1) = MD(t+1)M^T \tag{5.30}$$

Where **M** is transformation matrix, which is composed of three corresponding principal unite direction vector. All the principal damage components and their direction vectors together with current stress will be used to get the new anisotropic permeability based on Eq. 5.17-20.

As shown in Fig. 5.4, the working flow is generated based on the fundamental THM coupled framework introduced in chapter 3. Difference is that the damage-permeability models are integrated in this coupling. For the simulation of a HDR reservoir in this work, it begins at a model generation based on the specific geological data. The generated model is then discretizated in space. Together with some necessary input parameters, the simulation is been carried out on the THM coupled simulator, until times reaching at the time limit. Particularly, the simulation process is divided into two parts. In the first part, reservoir stimulation is simulated to create fracture network with valuable permeability for efficient heat extraction. In this stage, for a higher accuracy, a low time step of 200s is adopted during hydraulic fracturing, to overcome the uncertain brought by dramatic response against fluid injection. In the next stage, the heat extraction is conducted with a long time step up to 1×10^5s for efficient computation, based on the fracturing results containing some relevant information like geometry of created fracture network and permeability for the first stage. The data flow in this model is demonstrated in Fig. 5.5. Comparing with it in fundamental one (Fig. 3.16), the corresponding parameter consisted in damage as well permeability model is added in this new model.

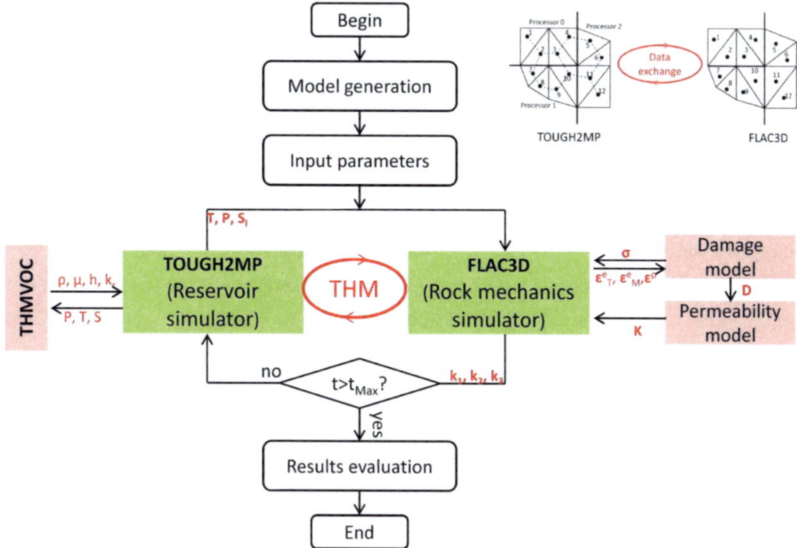

Figure 5.4. Flowchart of coupled THM model

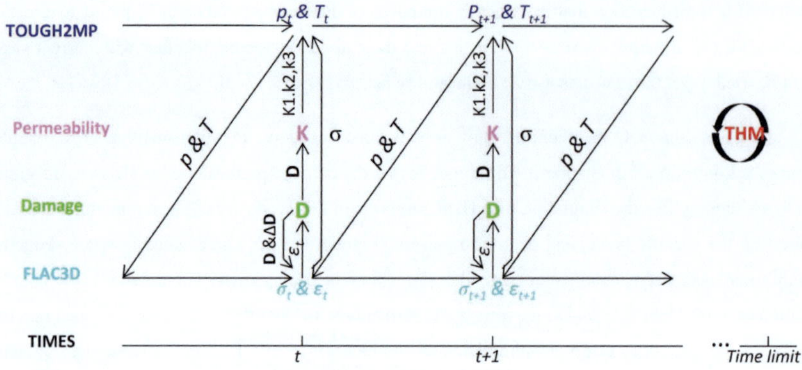

Figure 5.5. Flowchart of data in the developed model

5.4. Verification of implemented numerical model

The corresponding anisotropic damage model as well as derived damage- and stress-dependent anisotropic model has been implemented into THM coupled framework TOUGH2MP-FLAC3D. The implemented damage models and permeability model are verified separately by series of triaxial test on granite. The hydraulic fracturing is verified by small-scale flied hydraulic fracturing test.

5.4.1. Verification of implemented damage model

A simulation of a triaxial test on the granite from Beishan (Chen et al., 2015) was conducted under a confining pressure of 10MPa to verify this damage model. An isothermal condition with a room temperature of 25°C and a dry condition were assumed in this example. Tab. 5.1 summarizes all the used parameters. Except for the elastic parameters from Chen et al. (2015) and friction angle from Chen et al. (2014), the other parameters were obtained through matching the stress and damage during loading. Besides, the joint is not considered in this example, for the natural cracks were avoided as possible during preparation of the sample. As shown in Fig. 5.6a, the simulated axial stress and radial stress were plotted against the simulated axial strain, to compare the simulated results of Chen's model (Chen et al., 2015), which is a FEM based continuum damage model. The simulated axial stress curves from two models match very well at all tendencies such as peak strength as well as the rest strength. But the radial data of Chen's model is not given.

Additionally, the simulated averaged damage of the present model captures the mainly characteristics of limited experimental data and shows a comparable tendency with Chen's model (Fig. 5.6b). The damage is initiated at a very low rate. Once the axial stress closes or reaches the peak strength, the damage increases rapidly to a high level. After peak strength, the rate of damage slows down again and holds on a high level close to the upper limit.

Table 5.1. The used parameter in triaxial test of Beishan granite

Parameters	Value	
	Beishan Granite (Chen et al. 2015)	
Elastic parameters	E=70 [GPa]	v= 0.15 [-]
Plastic Parameters	$\varphi_m=\theta_m=35$ [°]	C=15 [MPa]
	σ_T =2 [MPa]	η_m = 2.6 [-]
	R=3.0 [-]	$b_1 = 2\times10^{-5}$ [-]
Damage parameters	a_1 =2 [GPa]	a_2 = -4.3 [GPa]
	r_0 =3×10^{-4} [-]	$r_1 = 5.5\times10^{-3}$ [-]
	D_{max} =1.2 [-]	k_d=0.5 [-]

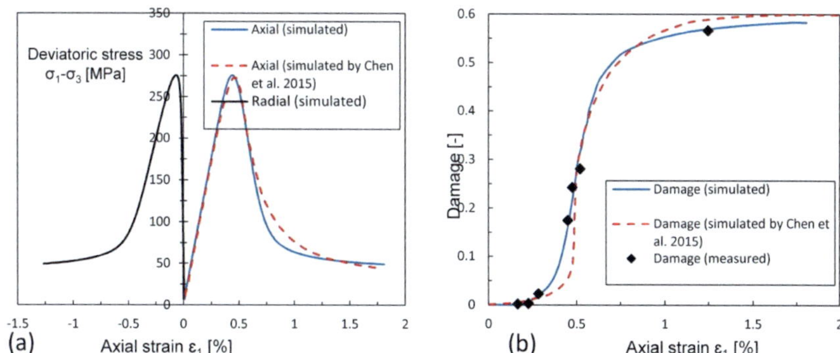

Figure 5.6. Comparing developed model with Chen at el. 2015 (a) stress-strain behavior; (b) strain-damage

5.4.2. Verification of implemented anisotropic permeability model

Let's consider the Eq. 5.16, which is function to determine the effective stress- and damage-dependent permeability. In order to check its plausibility, a triaxial test the granite from Lac du Bonnet has been carried out in this newly developed model. The testing condition and sample quality is same as the one mentioned above. Tab. 5.2 shows all the input parameters. After simulation, the achieved results, including strain-stress curve, stress-permeability and stress-damage are illustrated in Fig. 5.7, to compare with the measured results as well as the simulated results from Jiang et al. (2010). The simulated strain-stress behaviors show a very good agreement with the measured behavior, especially in axial direction. At beginning of loading, the simulated permeability decreases slightly, due to compressing under an increasing effective stress. Thereafter, the simulated permeability holds constantly, for the compacted sample cannot be compressed anymore. After initiating damage, simulated permeability will increase dramatically into a high level, along with growing damage. Overall, the simulated permeability captures the main characteristics of measured permeability as well as the simulated permeability by Jiang et al. (2010). In addition to this, the simulated permeability is also comparable with the simulated one.

Table 5.2. The used parameter in triaxial test of Lac du Bonnet granite

Parameters	Value	
	Lac du Bonnet granite (Souley et al. 2001)	
Elastic parameters	E=70 [GPa]	v= 0.28 [-]
Plastic Parameters	$\theta_m=\varphi_m$=46.4 [°]	C=15 [MPa]
	σ_T =2 [MPa]	η_m = 2.7 [-]
	R=3.0 [-]	b_1 = 2×10^{-5} [-]
Damage parameters	a_1 =2 [GPa]	a_2 = -4.3 [GPa]
	r_0 =1.1×10^{-3} [-]	r_1 = 1.0×10^{-2} [-]
	D_{max} =0.9 [-]	k_d =0.5 [-]
Hydraulic parameters	c_1=0.01 [-]	c_2=1.0×10^{6} [-]
	γ=0.25 [MPa^{-1}]	

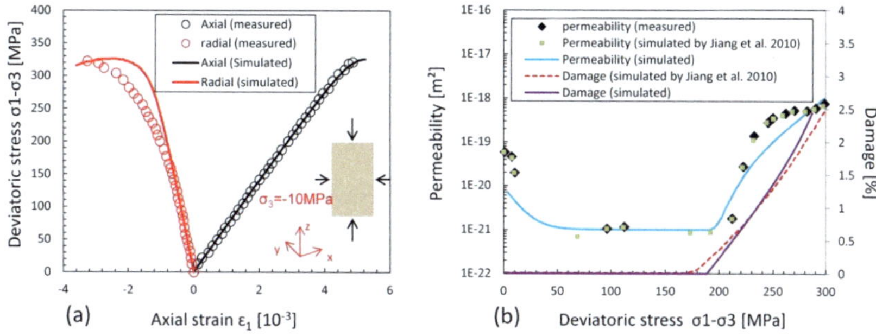

Figure 5.7. Comparing present model with results from Jiang at el. 2010 (a) stress-strain behavior; (b) stress-damage and stress-permeability

To verify the anisotropic permeability model, two triaxial tests on the granodiorite from Cerro Cristale and the granite from Westerly were simulated under a confining pressure of 10MPa and were compared with the simulated anisotropic permeability as well as measured permeability from Chen et al. (2014), in which the permeability of the REV is represented by the averaged permeability values of both matrix and microcracks over their respective volume fraction, according to the Viogt model. The permeability of microcracks is converted from the average aperture of the microcracks using the cubic law. Likewise, an isothermal condition with a room temperature of 25°C and a dry condition was used in this example. All used parameters are summarized in Tab. 5.3. Similar to that in example 1, no joints were considered. Final results are shown in Fig. 5.8 and 5.9. The axial and radial stress-strain curve of the present model matches the measured data much better, except the radial-strain curve of Westerly granite, in which the measured strain in radial direction is more significant (Fig. 5.8a and 5.9a). As shown in Fig. 5.8b and 5.9b, the simulated and measured permeability stay on a constant level, as the sample still in the elastic state. With the continuous loading, permeability increases gradually into two different values relating to axial and radial directions. The permeability in axial direction is higher than the one in radial direction. In Fig. 5.8b, the simulated permeability of this model matches well with the measured data, while it differs a little from the measured data as shown in Fig. 5.9b. In spite of this, their tendency and value are still comparable.

Table 5.3. The used parameter in triaxial test for anisotropic permeability

Parameters	Value			
	Cerro Cristale granodiorite (Chen et al. 2014)		Westerly granite (Chen et al. 2014)	
Elastic parameters	E=70 [GPa]	v= 0.28 [-]	E=70 [GPa]	v= 0.25 [-]
Plastic Parameters	$\theta_m=\varphi_m$=43 [°]	C=15 [MPa]	$\theta_m=\varphi_m$=43 [°]	C=15 [MPa]
	σ_T =2 [MPa]	η_m = 1.7 [-]	σ_T =2 [MPa]	η_m = 2.7 [-]
	R=1.8 [-]	$b_1 = 2\times10^{-5}$ [-]	R=1.8 [-]	$b_1 = 2\times10^{-5}$ [-]
Damage parameters	a_1 =2 [GPa]	a_2 = -4.3 [GPa]	a_1 =2 [GPa]	a_2 = -4.3 [GPa]
	r_0 =5×10^{-4} [-]	r_1 = 5.0×10^{-3} [-]	r_0 =6×10^{-4} [-]	r_1 = 5.5×10^{-3} [-]
	D_{max} =1.2 [-]	k_d =0.5 [-]	D_{max} =1.2 [-]	k_d =0.5 [-]
Hydraulic parameters	c_1=0.1 [-]	c_2=1.5×10^{5}[-]	c_1=0.1 [-]	c_2=1.5×10^{5}[-]
	γ=0.25 [MPa^{-1}]		γ=0.25 [MPa^{-1}]	

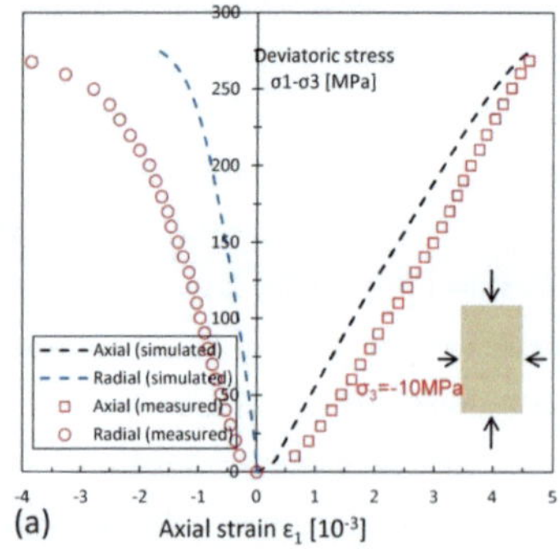

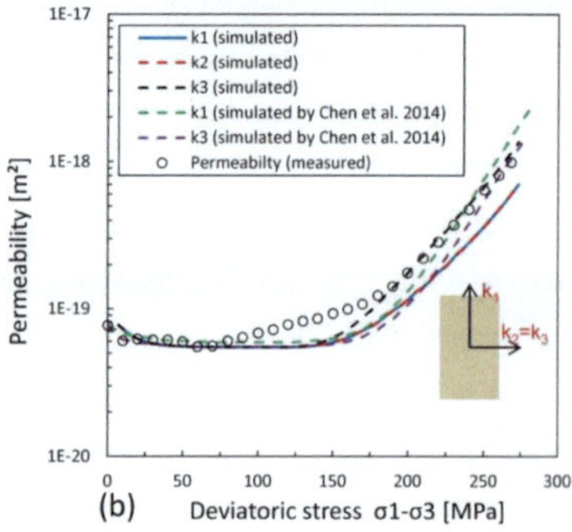

Figure 5.8. Comparing present model with results from Chen et al. 2014 (a) stress-strain behavior; (b) anisotropic permeability of Westerly granite

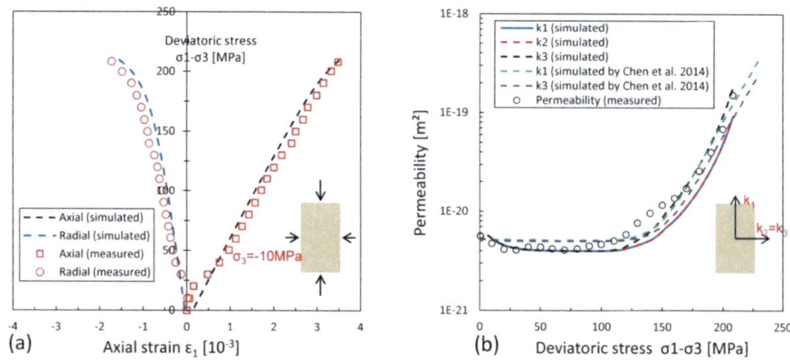

Figure 5.9. Comparing present model with results from Chen et al. 2014 (a) stress-strain behavior; (b) anisotropic permeability of Cerro Cristale granodiorite

5.4.3. Small scale fracturing test

Further to prove the plausibility of the developed model, the simulation of a small-scale hydraulic fracturing has been carried out based on the hydraulic fracturing experiments implemented at 410m depth in the Äspö Hard Rock Laboratory in Sweden. Wherein, the in situ hydraulic fracturing experiment on hard rock was performed in the tunnel TASN (Zang et al. 2016, Zimmermann et al. 2019). As shown in Fig. 5.10, the hydraulic fracturing test with six stages was performed in a horizontal borehole (F1). Three parallel horizontal monitoring boreholes M1-M3 was drilled for sensors installation. Many sensors and monitoring method were installed in surrounding tunnel. In this hydraulic fracturing test, the hydraulic fracturing with progressively increased cyclic injection treatment has been conducted and the accompanied seismic events have been measured in this procedure. Through analyzing the measured seismic events, the cyclic stimulation scheme is proven to have a low risk to induce seismic events in comparison with conventional hydraulic fracturing with a constant injection rate. Here, the hydraulic fracturing test NO.2 is selected to numerical study. Based on these data, a brick model is generated with the dimension of 80m×40m×40m, from which a quarter models is used to conduct the simulation. The initial vertical primary stress of 12 MPa is converted from the stress gradient along depth. Two horizontal stresses are initiated as 10MPa and 8MPa in x- and y-directions, respectively. A hydrostatic and isothermal condition is considered in this model. An isotropic permeability of $10^{-17}m^2$ and porosity of 0.1 is applied for whole the model. Other used parameters are concluded in Tab. 5.4. The simulation was carried out in stage separately. In flow back process, the flow back content is calculated by fixing the pressure at atmospheres pressure on the element, which locates in the borehole and connects directly with injection point.

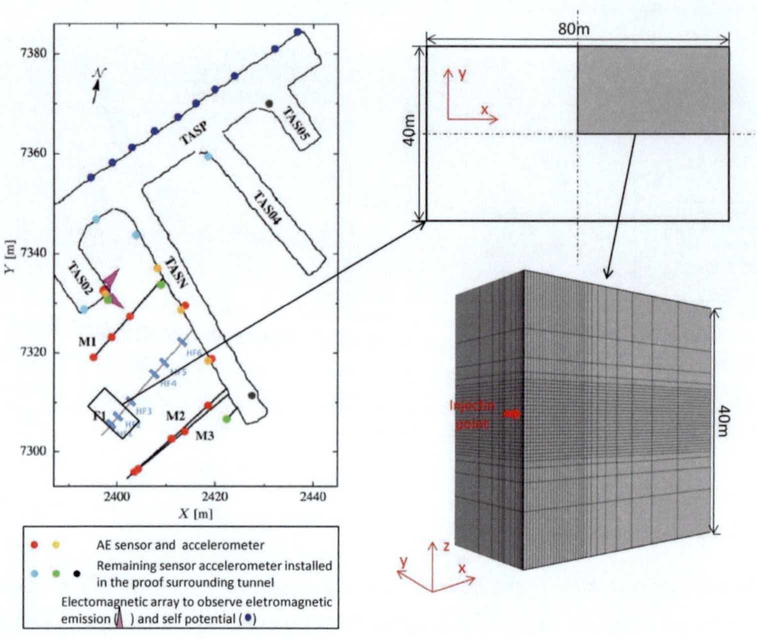

Figure 5.10. Location of Hydraulic fracturing test (Zimmermann et al. 2019) and numerical model

Table 5.4. The used parameter for hard rock fracturing

Parameters	Value	
Elastic parameters	E=50 [GPa]	v= 0.20 [-]
Plastic Parameters	$\theta_m=\varphi_m$=40 [°]	C=4 [MPa]
	σ_T =1 [MPa]	η_m = 2.0 [-]
	R=3.0 [-]	$b_1 = 2\times10^{-5}$ [-]
Damage parameters	a_1 =2 [GPa]	a_2 = -4.3 [GPa]
	r_0 =1.2×10^{-4} [-]	$r_1 = 2.0\times10^{-3}$ [-]
	D_{max} =1.2 [-]	k_d =0.5 [-]
Hydraulic parameters	c_1=0.01 [-]	c_2=1.0×10^7 [-]
	γ=0.5 [MPa^{-1}]	

During fracturing and flow back, the pressure and permeability at injection element are recorded and are plotted together with the in situ measured pressure and permeability in Fig. 5.11. Clearly, the simulated injection pressure shows a very similar tendency and captures most characteristics of measured injection pressure. One exception that the simulated injection pressure in the first cycle is obviously exceeding the measured one, for the reason that the pre-fracturing in situ is not considered in this simulation, thus, the breakdown pressure in virgin rock could reach very high. As shown in Fig. 5.11b, the simulated permeability at injection point and in-situ measured permeability are compared. The permeability is enhanced by hydraulic fracturing and shows an increasing value, for the damage is accumulated continuously with fracturing cycles. In first three cycles, both permeability match very well. Then, the simulated permeability increases continuously and separates gradually from the measured one, which only has a slight change, as the damage in the newly developed model is irreversible. In spite of this, the permeability from present model and field experiment is still comparable.

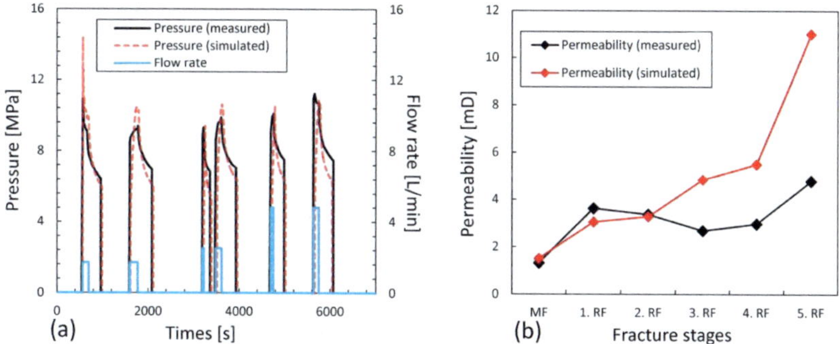

Figure 5.11. Comparing the simulated and measured (a) injection pressure; (b) permeability at injection point

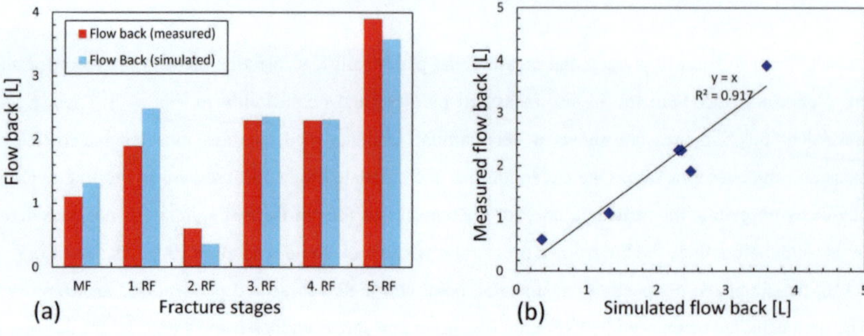

Figure 5.12. Comparing the simulated and measured flow back content

After the injection in each cycle fracturing, pumping is stopped for a while. Then, the flow back is carried out until starting next cycle fracturing. The flow back contents accumulated in each cycle fracturing are summarized in Fig. 5.12a together with in-situ measured content. The simulated flow back content relates to the duration of flow back as well as the pore pressure. To compare the results much visibly, simulated and measured flow back contents are plotted respectively as x- and y-coordination in Fig. 5.12b. Clearly, a linear correlation between simulated and measured results with a correlation coefficient of 0.917 verifies the feasibility of the newly developed model. Additionally, the simulated fracture profile is superimposed on the measured seismic events during fracturing in Fig. 5.13. From the over view in Fig. 5.13a, it can be found that the measured distributed on one side of injection well and along the well, for the probable reason that some drilling triggered damages could change the propagation of fracture. Such characteristics are also found in the side view in Fig. 5.13b. In spite of this, if we focus on half wing of fracture, the simulated fracture could capture most principal characteristic of in situ stimulated fracture.

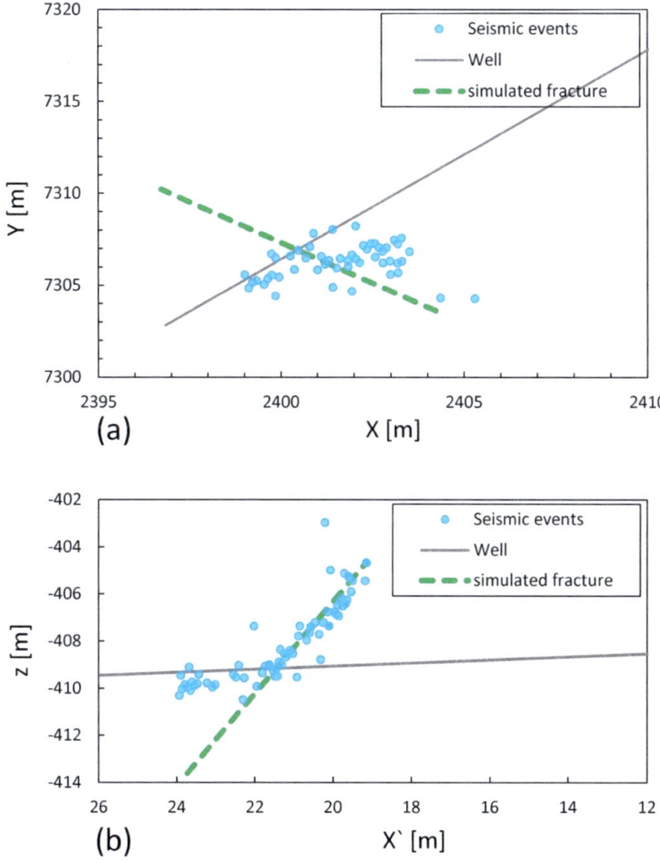

Figure 5.13. Comparing the simulated fracture with seismic events recorded during fracturing test (a) over view; (b) side view

5.5. Influence factors on reservoir stimulation

A 2D square model with unit thickness and side lengths of 100m is discretized into 10,000 square elements and used to study the influence of injection rate, reservoir temperature and injection temperature on stimulation in geothermal reservoir. As demonstrated in Fig. 5.14a, CO_2 was injected at the bottom right corner to create a fracture network. The plastic parameters on joint are accounted as only 80% of them being on the matrix, which together with the elastic parameters are shown in Fig. 5.14a. The rest of the damage and permeability parameters are taken from granite as presented in Tab.

5.3. The variables, including the injection fluid, mass rate, reservoir temperature and injection temperature in different cases are given in Tab. 5.5. 1% CO_2 and 99% water constitute the initial composition in the geothermal reservoir. After 1 hour injection of CO_2 in case 1, a fracture network is illustrated by the pressure contour in Fig. 5.14b.

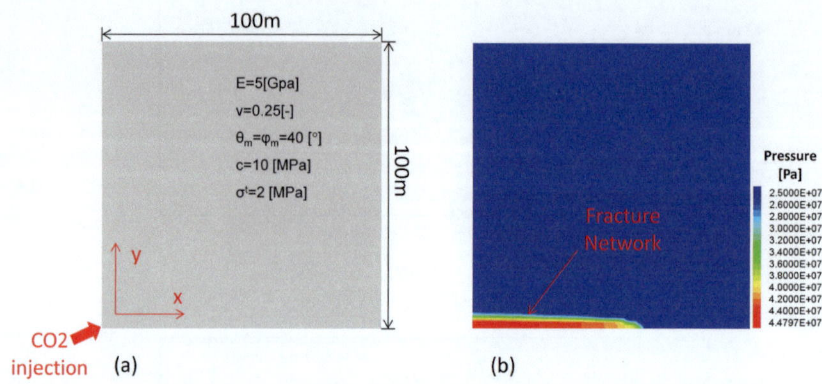

Figure 5.14. (a) geometry and some used parameters of the model; (b) the fracture network after 1 hour injection

Table 5.5. The used parameter for hard rock fracturing

Case	Injection fluid	Injection rate [kg/min]	Injection temperature [°C]	Reservoir temperature [°C]
1	CO_2	2kg/min	40°C	150°C
2	CO_2	1kg/min	40°C	150°C
3	H_2O	2kg/min	40°C	150°C
4	H_2O	1kg/min	40°C	150°C
5	CO_2	2kg/min	40°C	100°C
6	CO_2	2kg/min	40°C	180°C
7	CO_2	2kg/min	60°C	150°C

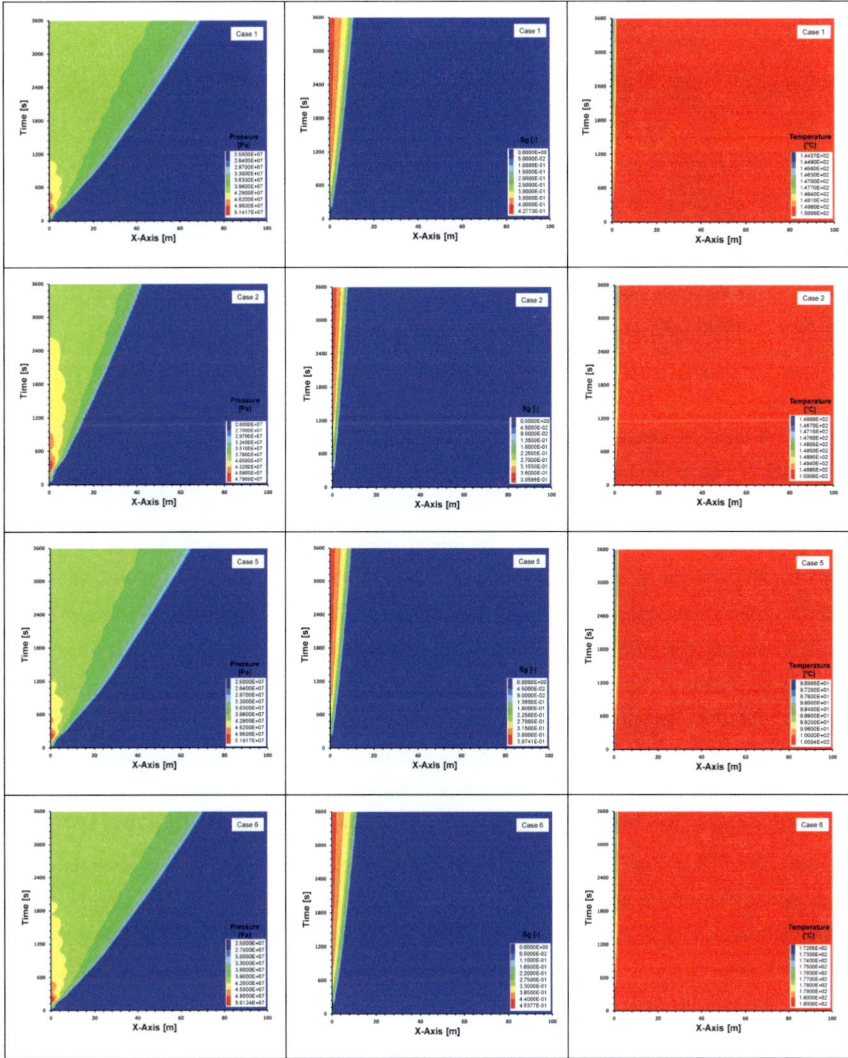

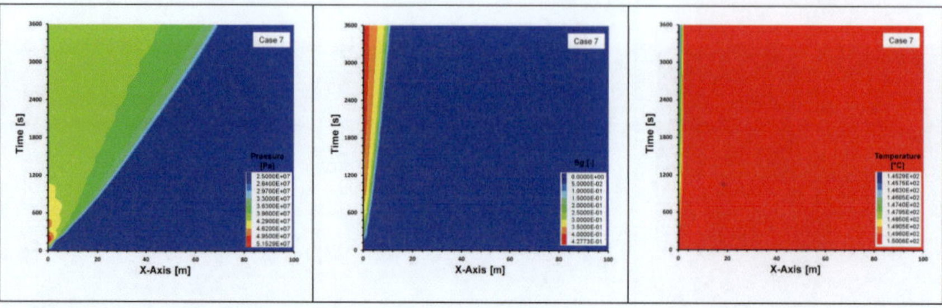

Figure 5.15. The temporal evolution of pressure, gas saturation and temperature along the line crossing injection point in x direction

In all CO_2 injection cases, the temporal evolution of pressure, gas saturation and temperature are compared in Fig. 5.15 along the line crossing the injection point in x direction. Generally, the proceeding speed of CO_2 is slower than fracture network propagation, because the pressurized water by injected CO_2 in fracture tip can derive the fracture propagation. In comparison with the high heat capacity of reservoir rock, the cold CO_2 can only reduce the reservoir temperature a few grades at the injection point. On the other hand, in case 1 and 2, the high injection rate promotes the propagation of fracture network. Correspondingly, the maximum pressure in case 1 is higher than that in case 2. There is no significant difference in case 1 and case 7, when the injection temperature is increased from 40°C to 60°C. With the reservoir temperature rising from 100°C to 150°C, the fracture length has a noticeable increase in case 1 and 5. And the increase can be ignored in case 1 and 6 with reservoir temperature ranging from 150°C to 180°C, whereas the pressure shows an inverse relation with the reservoir temperature in these cases. Such tendencies can be explained by observing the state properties of CO_2 in Fig. 5.16a. Viscosity and density of CO_2 have an inverse relationship with temperature. Under the same mass higher density means less volume. Therefore, less volume of CO_2 with high viscosity creates a shorter fracture under high pressure in case 5, which is proved by the saturation contours as well. The CO_2 invaded depth and saturation in case 5 are lower than that in case 1 and 6. Additionally, the state properties of CO_2 in case 1 are illustrated also at different times in Fig. 5.16b. The pressure at the start point (injection point) rises to a high level with high viscosity and density during CO_2 injection. Namely, the CO_2 is compressed at the injection point. And the state properties of CO_2 are dynamic with mutating pressure and temperature conditions.

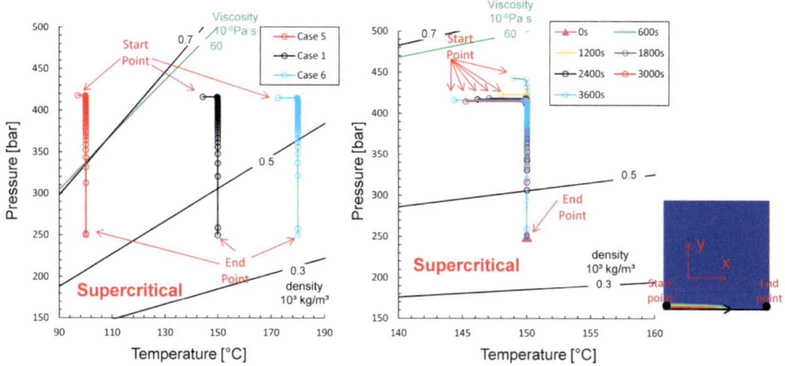

Figure 5.16. pressure versus temperature from the start point (injection point) to end point (a) in case 1, 5 and 6 after 1 hour fracturing, (b) in case 1 at different time points

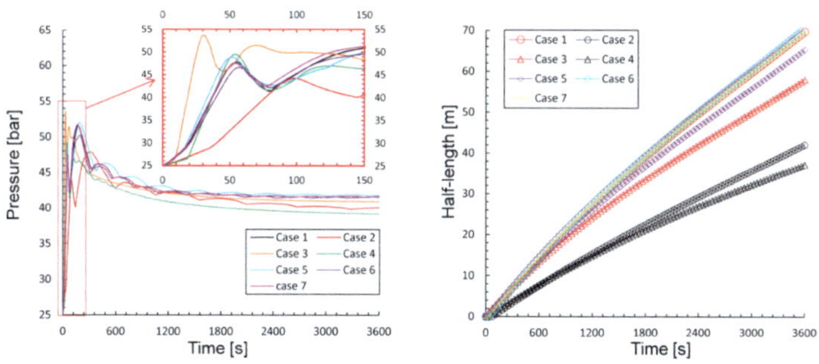

Figure 5.17. comparing (a) pressure at injection point, and (b) fracture half-length of different cases

The pressure curve at injection point of all cases is plotted against the injection time in Fig. 5.17a. Similar to the above discussed results, the break down pressure shows a direct relation with injection rate, while an inverse relation with reservoir temperature. Due to the pressure- and temperature-dependent expansivity, the pressure of CO_2 injection in case 1 grows much slowly with a lower break down pressure, compared with water injection in case 3. Such tendency is shown in the compared cases 2 and 4 as well. Especially, the pressure of CO_2 injection drops in the form of a wave after reaching the maximum value, which is caused by the compressibility of CO_2 and discrete model. However, the pressure curve of water injection evolves more smoothly at a lower pressure level after fracture initiation. The final fracture half length is shown in Fig. 5.17b. By comparing the groups of cases 1-3 and 2-4, CO_2 injection can create a longer fracture than that of water, owing to the relatively lower density of CO_2 around 500kg/m³ i.e. a large injected volume in this condition.

5.6. Description of planned EGS project in Dikili

Dikili ESG is the first EGS project in Turkey, which is being planned at license area of SDS Energy Inc., in Dikili of the İzmir province. As shown in Fig. 5.18, the candidate site locates on Anatolia bordered by the Mediterranean Sea, where the transitional zone of African plate and Eurasian plate. The geological tectonic activities are active strongly in this region, as a results, the thermal gradient of about 7°C/100m in the candidate site is remarkable (Hou et al. 2015b), in comparison with general gradient of 3°C/100m. Besides, many hot springs have been found around target region. In the light all of these, this candidate site has high potential for implementing the geothermal project. And the two candidate areas (area A and B) are assigned in the license area. In this thesis, the hydraulic fracturing as well as heat production is conducted based on area A.

Figure 5.18. Location of planned Dikili geothermal project

In this regions, extensive geological, paleostress (279 fault-slip data from 33 locations), geophysical (magnetotelluric and vertical electrical sounding at 80 and 129 locations, respectively) and geochemical studies as well as paleostress measurements have been conducted in this area. Some important geological information is illustrated in Fig. 5.19. In the license area, the direction of minor stress is found dominantly in north-south direction. The cross-section along orange line (in Fig. 5.19a) is illustrated in Fig. 5.19b. The more detail strata information, including geology and rock type is given in Fig. 5.20. The target geothermal reservoir is determined in Yuntdagi volcanic, in which the naturally fractured and altered zone maybe provides some potential channel for heat production. The deeper formation-Kozak Pluton with grnodioritic magmatic rocks is also treated as geothermal reservoir. The thick tuff and marl layers of the Yuntdagi Formation, including the marl and limestone units of the Ularca Formation, which unconformably overlie the Yuntdagi Formation, provide the caprock to the reservoir. However, the natural permeability of target geothermal reservoir is still not enough for an efficient heat production. Hydraulic fracturing need be conducted firstly to enhance the permeability before heat production.

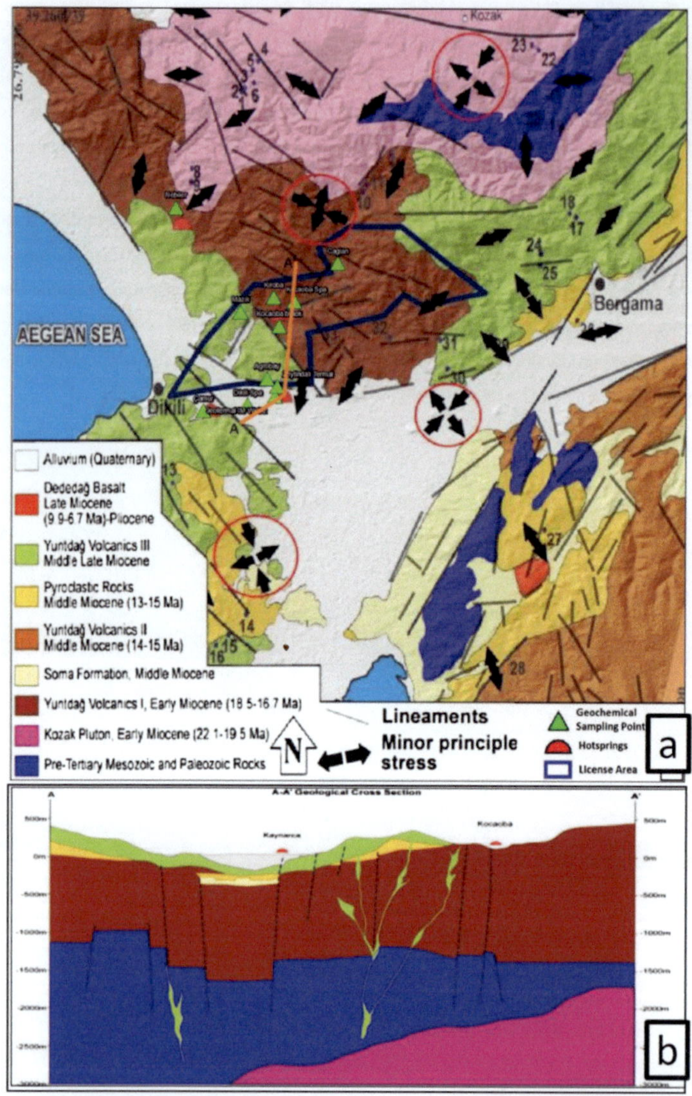

Figure 5.19. Geological map of the region and stress orientations, (b) Cross-section of the orange line A-A' in (a) (modified from Liao et al. 2020b)

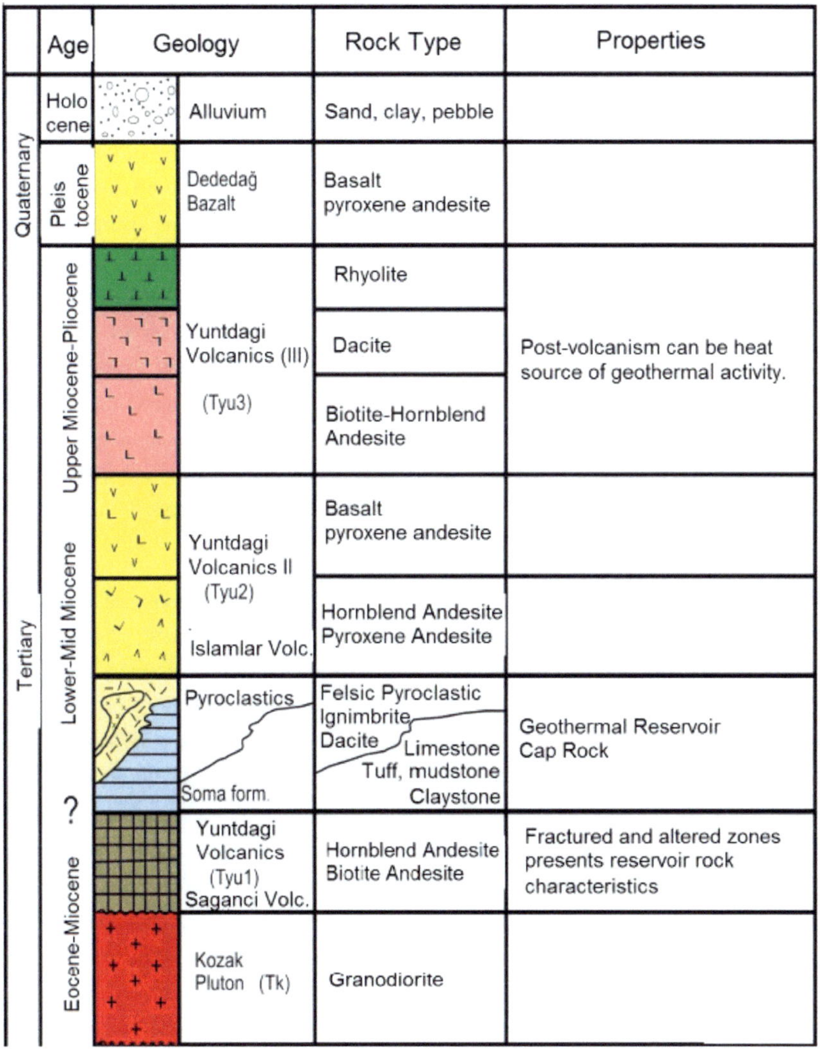

Age		Geology	Rock Type	Properties
Quaternary	Holocene	Alluvium	Sand, clay, pebble	
Quaternary	Pleistocene	Dededağ Bazalt	Basalt pyroxene andesite	
Tertiary — Upper Miocene-Pliocene		Yuntdagi Volcanics (III) (Tyu3)	Rhyolite	Post-volcanism can be heat source of geothermal activity.
Tertiary — Upper Miocene-Pliocene		Yuntdagi Volcanics (III) (Tyu3)	Dacite	Post-volcanism can be heat source of geothermal activity.
Tertiary — Upper Miocene-Pliocene		Yuntdagi Volcanics (III) (Tyu3)	Biotite-Hornblend Andesite	Post-volcanism can be heat source of geothermal activity.
Tertiary — Lower-Mid Miocene		Yuntdagi Volcanics II (Tyu2)	Basalt pyroxene andesite	
Tertiary — Lower-Mid Miocene		Islamlar Volc.	Hornblend Andesite Pyroxene Andesite	
Tertiary — Lower-Mid Miocene		Pyroclastics / Soma form.	Felsic Pyroclastic Ignimbrite Dacite / Limestone Tuff, mudstone Claystone	Geothermal Reservoir Cap Rock
Tertiary — Eocene-Miocene (?)		Yuntdagi Volcanics (Tyu1) Saganci Volc.	Hornblend Andesite Biotite Andesite	Fractured and altered zones presents reservoir rock characteristics
Tertiary — Eocene-Miocene		Kozak Pluton (Tk)	Granodiorite	

Figure 5.20. geological information and rock types at candidate site (SDS Energy 2014)

5.7. Numerical model of Dkili EGS and initial condition

The reservoir stimulation, heat extraction as well as CO$_2$ sequestration in a CO$_2$ based geothermal reservoir with pre-existing formation water was studied comparing with traditional working fluid, water. As shown in Fig. 5.21, a simplified 3D model was built on the base of geological information and stratigraphy from the planned EGS in Dikili, Turkey. This geothermal reservoir composed of Kozak Pluton formation with dominant granite lies at the depth from -2860 m to -1860 m and is covered by the thick tuff and marl layers of the Yountdagi formation, and the marl and limestone units of the Ularca Formation at the top (Hou et al. 2015b). At the bottom of geothermal reservoir, a 640 m thick base layer is taken accounted in this model. Virgin geothermal reservoir is occupied almost by water. Initial pore pressure converted from the hydraulic gradient of 0.01MPa/m. The applied initial principal stresses are plotted against the true vertical depth in Fig. 5.21 as well. According to Hou et al. (2015), a remarkable temperature gradient of 0.07 °C/m is expected and used to calculate formation temperature in candidate site of projected geothermal reservoir. Here, only a quarter models with dimensions of 2000 m×1000 m×2500 m is considered in the simulation, wherein a pay zone with a thickness of 100 m in geothermal reservoir is selected for reservoir stimulation and heat extraction. This simulation model is discretized into 74,700 rectangular elements. An injection point and production point are contained in this model at the depth ranging -2400~-2500m and -2200~-2300m respectively. The production point posits higher than injection point. They are separated by a distance of 1000 m from each other in horizontal direction.

In this model, the rock properties are summarized in Tab. 5.6 separately according to reservoir formation. Due to the similar dominating rock types, the Ularca and base formation adopts the same value as that used in Kozak Pluton and Youndtagi formation respectively. Referring to the high friction angle of granite (Chen et al. 2014), the friction and dilation angle of 40° is used in all formation. The initial isotropic permeability of 4×10^{-18} m² is applied Kozak Pluton and base formation (Hou et al. 2015b). In both of the cap formations, the initial permeability is 1×10^{-19} m², thereby γ is 1.2×10^5 and 5×10^6 in Ularca and Youndatagi formation, Kozak Pluton formation, respectively, which is calculated from the maximum permeability 5×10^{-13}m² referring to Liao et al. (2019b). The rest plastic parameters and damage parameters are taken from the triaxial test matching of granite in Tab. 5.3. In geothermal pay zone, an anisotropic natural joint is considered with the dominant orientation in y-direction. The plastic parameters of joint, including frictions angle, dilation angle, cohesion, and tensile strength, are assumed only 80% of the value applied on the matrix. An identical thermal expansion of 1×10^{-6} [1/°C] is used in whole model. For reservoir thermal characteristics, the thermal conductivity and specific heat capacity were assumed as 2.83 W/(m K) and 965 J/(kg K), respectively.

To create an efficient fracture network, total 90,000 kg CO_2 or water with a temperature of 40 °C is injected into geothermal reservoir. As shown in Fig. 5.22, the injection begins at a low injection rate of 37.5 kg/s and increases stepwise to 150kg/s in the first 0.45 hours (h), followed by maintaining constant injection rate of 150 kg/s. After that, the injection rate is firstly reduced to 75kg/s at 160h and to 37.5 kg/s at 170h. The injection is terminated at 180 h and the simulation continues till 250 h.

Table 5.6. Input parameters in the application

Parameters		Value				References
		Ularca Formation	Youndtagi Formation	Kozak Pluton Formation	Base Formation	
Elastic parameters	$E[GPa]$	62	62	55	55	Hou et al. 2015b
	$v[-]$	0.23	0.23	0.20	0.20	Hou et al. 2015b
Plastic Parameters	$\theta_m=\varphi_m[°]$	40	40	40	40	Chen et al. 2014
	$c[MPa]$	20	20	10	10	Hou et al. 2015b
	$\sigma_T[MPa]$	6	6	4.6	4.6	Hou et al. 2015b
	$\eta_m[-]$	2.7	2.7	2.7	2.7	Granite in Tab 5.3
	$R[-]$	1.8	1.8	1.8	1.8	Granite in Tab. 5.3
	$b_1[-]$	2×10^{-5}	2×10^{-5}	2×10^{-5}	2×10^{-5}	Granite in Tab. 5.3
Hydraulic parameters	$\gamma[MPa^{-1}]$	0.25	0.25	0.25	0.25	Granite in Tab. 5.3
	$c_1[-]$	0.1	0.1	0.1	0.1	Granite in Tab. 5.3
	$c_2[-]$	1.2×10^{5}	1.2×10^{5}	5×10^{6}	5×10^{6}	Maleki & Pouya 2010
	$K_0[m^2]$	4×10^{-18}	4×10^{-18}	1×10^{-19}	1×10^{-19}	Hou et al. 2015b
	$\phi[\%]$	1.29	1.29	2	2	Hou et al. 2015b
Damage parameters	$a_1[GPa]$	1	1	1	1	Granite in Tab.5.3
	$a_2[GPa]$	-2.3	-2.3	-2.3	-2.3	Granite in Tab. 5.3
	$r_0[-]$	2×10^{-4}	2×10^{-4}	2×10^{-4}	2×10^{-4}	Granite in Tab. 5.3
	$r_1[-]$	5.5×10^{-3}	5.5×10^{-3}	5.5×10^{-3}	5.5×10^{-3}	Granite in Tab. 5.3
	$k_d[-]$	0.5	0.5	0.5	0.5	Granite in Tab. 5.3
	$D_{max}[-]$	1.2	1.2	1.2	1.2	Granite in Tab. 5.3
Thermal parameter	$\beta[1/°C]$	1×10^{-6}	1×10^{-6}	1×10^{-6}	1×10^{-6}	Hou et al. 2015b

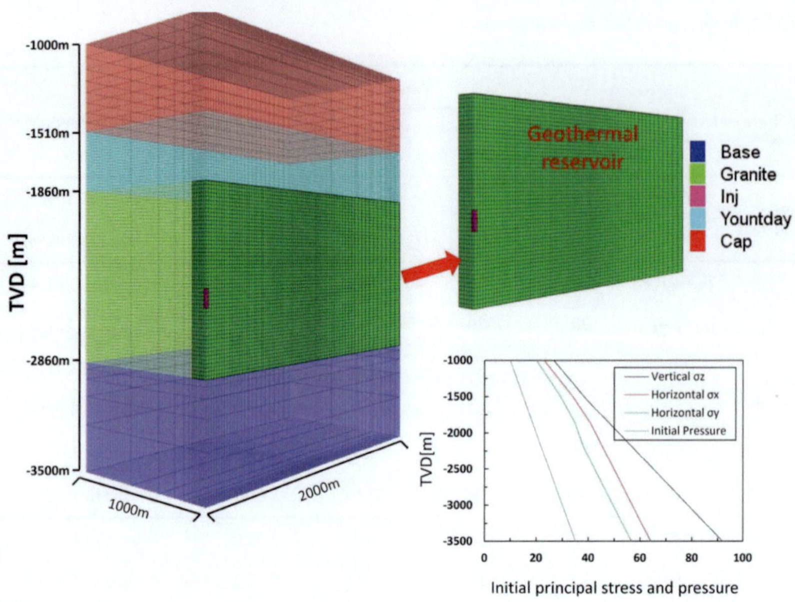

Figure 5.21. Simplified 3D geological model for simulation and initial principal stress (Liao et al. 2019b)

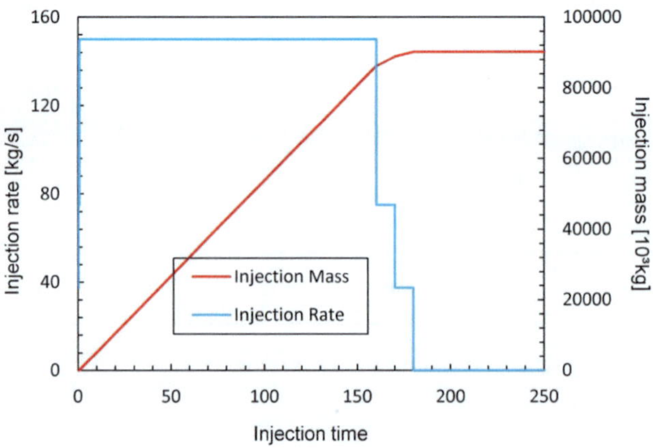

Figure 5.22. hydraulic fracturing treatment for both water as well as CO_2 injection

5.8. Reservoir stimulation using supercritical CO_2

Finally, a fracture network with maximum half length about 1200 m is created by CO_2 injection, in which the corresponding permeability in x-, y- and z-directions are shown in logarithmic form in Fig. 5.23a-c respectively. The permeability in x- and z- direction is noteworthy compared to y-direction, since the fracture propagates perpendicular to minimum principal stress along the dominant joint. In x-direction, the permeability contour spreads in a radial shape from the injection point and is limited in the pay zone with a height of approximate 1000 m. The permeability decreases away from the injection point except for the small zone near to it. The maximum permeability of 1.78×10^{-13} m² is located at the top right about 200 m from injection point. The small zone near the injection point shows a relatively low permeability in x- and z-directions, resulting from that the deformation perpendicular to x-z plane is limited by the relative high deformation perpendicular to y-z plane around injection point because of a non-zero Poisson ratio. And the relative high deformation perpendicular to y-z plane enhances the permeability in y-direction (see in Fig. 5.23b). The permeability in z-direction shows a very similar distribution shape with the maximum value of 1.78×10^{-13} m². The maximum permeability in y-direction having a value of only 1.78×10^{-14} m² is found at injection point, which is about ten times smaller than that in x- and z-directions, forming a fracture network propagation with the maximum width of 40 m in y-direction. In comparison with CO_2 injection, a very analogical shape of permeability distribution is created by water injection (see in Fig. 5.23d-f). However, the fracture networks created by water and CO_2 are obviously different. The half length of the fracture network created by water injection is only 1030 m, which is nearly 200 m less than that of CO_2 injection. In addition, the maximum permeability of 1.35×10^{-13} m², 2.71×10^{-14} m² and 1.36×10^{-13} m² in water-based fluid injection created fracture network is slightly lower than that in CO_2-based fluid injection created fracture network, in x, y and z-direction respectively.

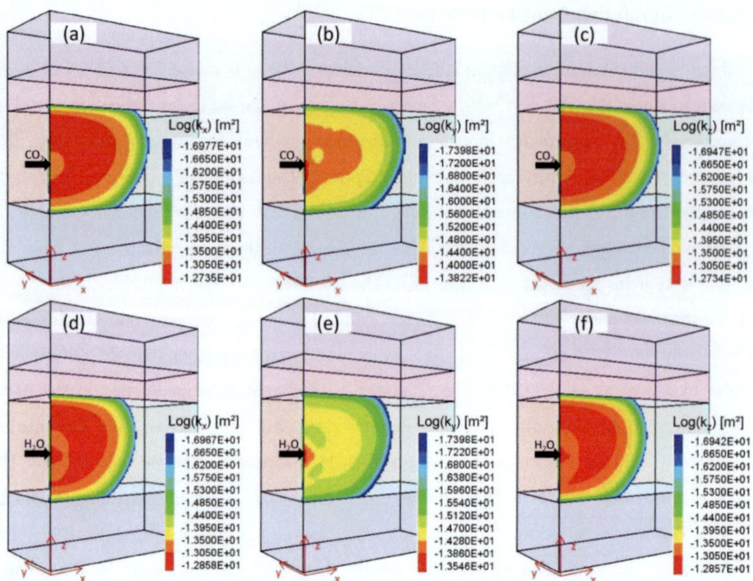

Figure 5.23. The permeability distribution of the fracture networks created by water and CO_2 respectively (Liao et al. 2019b)

After fracturing, the pore pressure distribution of CO_2 injection at 250 h is shown in Fig. 5.24a. Understandably, the pore pressure in created fracture network is higher than that in its surroundings. And the maximum pore pressure of 48.7 MPa is found at the vicinity of injection point. During both water and CO_2 injection, the injection pressure at the depth of -2450m was recorded against the injection time in Fig. 5.24b. The fracturing pressure of CO_2 injection is about 73 MPa, which is almost 17 MPa lower than that of water injection. This demonstrates the great superiority of CO_2 on fracturing in comparison with water, which is in according with the experimental results of Wang et al. (2018). After the creation of fracture, the injection pressure increases continuously and exceeds the water injection pressure. The injection pressure reaches the peak value of 98 MPa in 2.7 h during CO_2 injection, and then drops quickly below to that of water injection. This is result from that only a lower pressure is needed to keep the fracture propagating. In addition, the viscosity of CO_2 is far lower comparing with water, leading to a lower injection pressure needed. When injection rate reduces from 150kg/s to 75kg/s at 160h, the injection pressure of both CO_2 and water injection drops instantly to relatively low value. Moreover, owing to the expansivity of CO_2, the injection pressure drops at a much slower rate. With further reduction of injection rate to 37.5 kg/s at 170 h, the pressure of CO_2 injection exceeds that of water again, this is also result from the higher expansivity of CO_2. After the discontinuation of pumping, the injection pressure reduces gradually to 45.7 MPa and 43 MPa for CO_2 and water injection respectively.

The stimulated reservoir volume (SRV) and area (SRA) by both CO_2 and water injection are plotted against the injection time in Fig. 5.25a. The SRV and SRA evolve positively with injection time and slow down after pumping is stopped. Finally, the SRV and SRA reached at 8.7×10^7 m³ and 2.2×10^6 m² with CO_2 injection, and at 7.9×10^7 m³ and 1.9×10^6 m² with water injection respectively. The average width of 38.9 m and 41.5 m are converted from corresponding SRV and SRA of CO_2 and water injection respectively. The maximum half length corresponding to CO_2 and water injection is shown in Fig. 5.25b as well. In the beginning, the two fracture lengths are almost overlapping and increase rapidly with a steep slope, as the fracture is not fully extended in vertical directions. When the fracture has reached its height limit, fracture lengths grows at a constant rate. Whereas, the rate of length growth during water injection is slower than that of CO_2 injection, for the reason that the pressure driven leak off is more significant for water injection (see in Fig. 5.24b). on the other hand, the expansion of CO_2 can further enhances the stimulated fracture network.

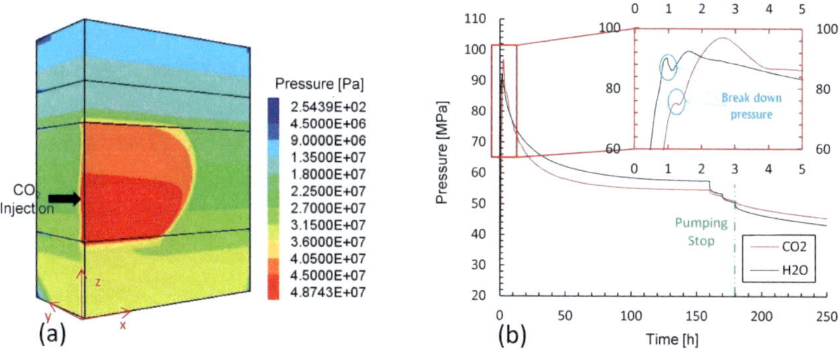

Figure 5.24. (a) the created fracture network; (b) the bottom hole pressure of water and CO_2 injection(Liao et al. 2019b)

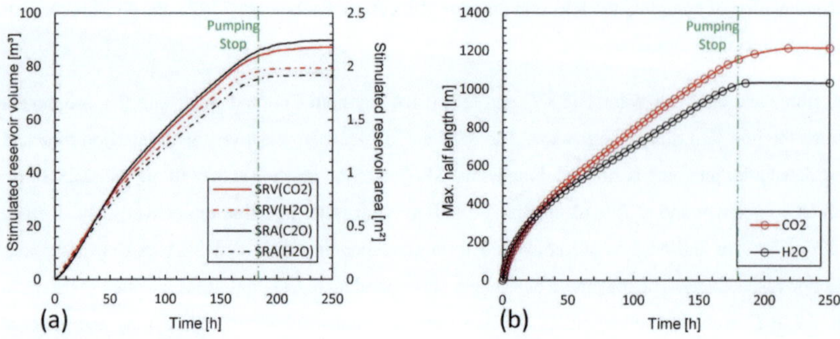

Figure 5.25. Comparing (a) stimulated fracture volume and area (b) fracture half-length of water and CO_2 created fracture network (Liao et al. 2019b)

5.9. Heat production using supercritical CO_2

Based on the created fracture networks, the heat extraction over 30 years is conducted with continue CO_2 and water injection at rate of 100 kg/s, respectively. The production pressure is fixed at 21.1 MPa, which is about 0.5 MPa higher than the hydrostatic pressure at the production point. The gas saturation and temperature contours of CO_2 and water injection at 250 h, 10 years (a), 20a and 30a are shown in Fig. 5.26. At 250 h (after fracturing), the CO_2 is concentrated in a semicircular region with the gas saturation below 0.5. This semicircular region is much smaller than fracture network (see in Fig. 5.23a). The temperature decreases only in the vicinity of injection point. With continuous injection and production, the CO_2 spreads gradually to the production point. At 10 a, the CO_2 tip has already reached the production point. And the reservoir can be clearly categorized into three zones, including the central, intermediate and peripheral zone. The CO_2 is mainly concentrated in the central zone with gas saturation close to 1. Due to the significantly lower viscosity compared with water, the hot CO_2 meeting lower resistance will pierce formation water to form a priority channel from injection point to production point, which is beneficial for heat extraction. This priority channel effect is also confirmed by the temperature contours. The temperature drops substantially in the priority channel. And then the priority channel spreads outwards along with production time. Most of heat is mined in the priority channel with lowest temperature measured equal to 60°C (injection temperature). However, the heat is mined in a much wider area by water injection, because, comparing with the high viscosity difference between water and CO_2, similar viscosity makes the injected water spread much uniformly into surrounding.

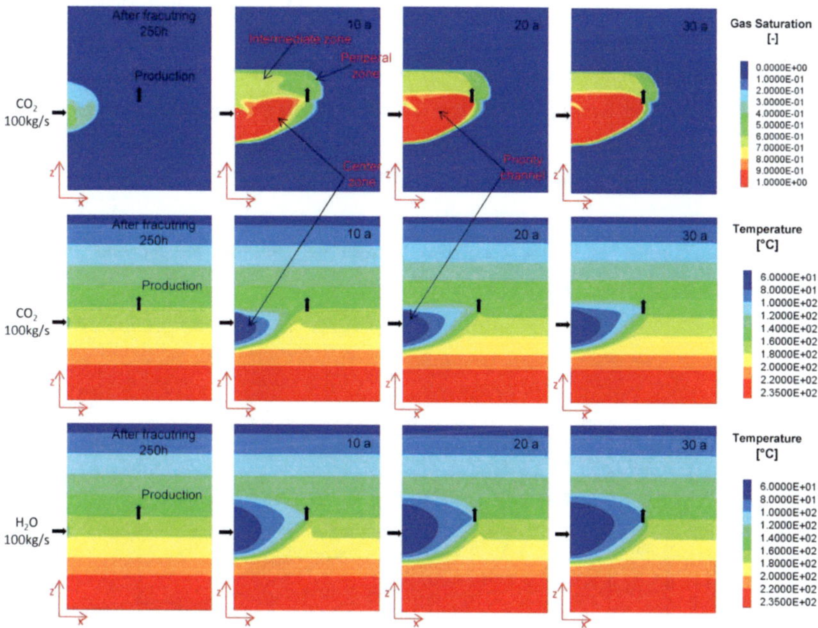

Figure 5.26. Temperature and gas saturation distribution during production driven by water and CO_2 respectively (Liao et al. 2019b)

Furthermore, to study the influence of injection rate on the heat production, the simulations with CO_2 and water injection at 60kg/s, 80kg/s and 120kg/s were conducted respectively. Fig. 5.27 shows the average production temperature as a result of different rates for CO_2 and water injection. The temperature of CO_2 injection falls a few degrees during the first several months, for the reason that the buoyancy due to lower density of CO_2 accelerates the fluid flowing above injection point. Therein the fluid has lower temperature. While, as shown in Fig. 5.28, the high vertical pressure gradient of water injection, especially the obvious pressure increment at injection point promotes fluid flow from injection point to production point. As there is about 200 m vertical difference between injection point and production point, the fluid from deeper formation causes a slight increase of production temperature at beginning. This behavior is illustrated by front view of flow vector in temperature contours after 100 days production in Fig. 5.29. In the case of CO_2 injection, the flow from top-left corner of production point is relative strong as the one by water injection at the same place, whereas the flow due to water injection from lower left is more significant comparing with CO_2 injection (see in the red circle). With continuous water injection, the production temperature of water injection decreases gradually after a few years because of heat mining. However, the production temperature of CO_2 injection has a minor increase, this is result from that the injection point and production point has been connected by priority channel, which is beneficial for bringing more fluid with higher temperature from deeper formation. After that, production temperature decreases gradually as well. Generally, in the first few years, production temperature of CO_2 injection is lower than the one of water, but with continuous production, the production temperature of CO_2 injection gradually exceeds it of water injection, because, owing to the high specific heat capacity, water is more efficient for heat extraction. Additionally, an inverse correlation between the injection rate and end production temperature can be concluded in both water and CO_2 injection.

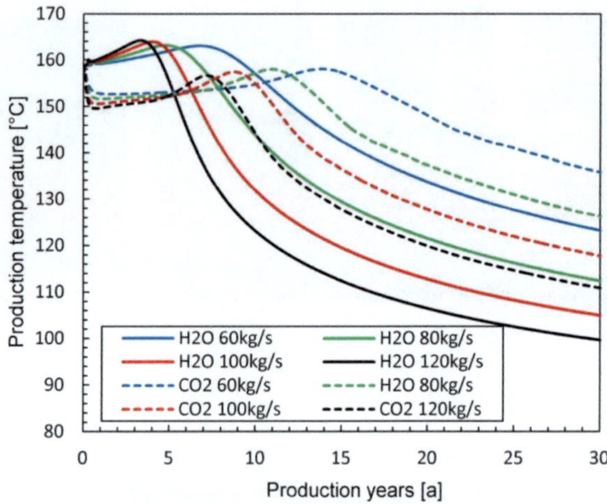

Figure 5.27. The temperature of produced fluid at different injection rates (Liao et al. 2019b)

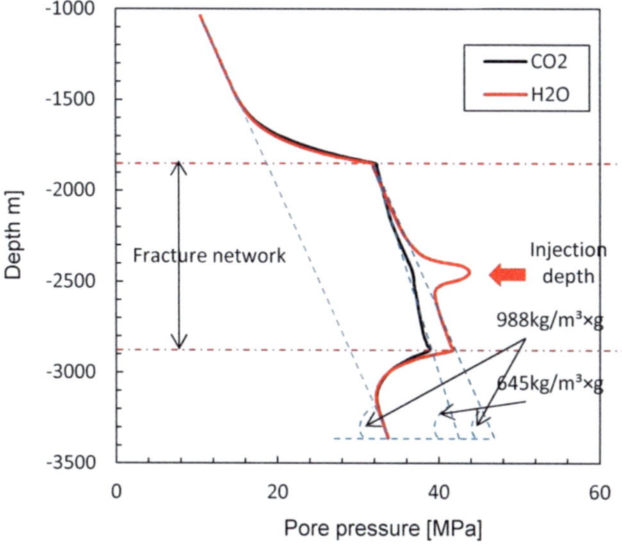

Figure 5.28. The pressure distribution along the vertical straight line crossing injection point(Liao et al. 2019b)

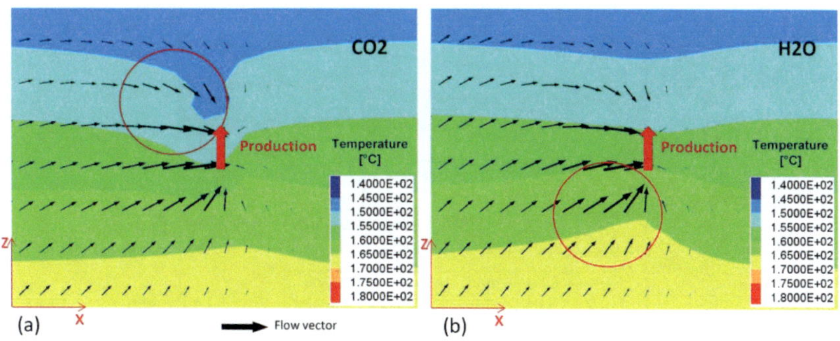

Figure 5.29. The flux surrounding production point after 1 year production(Liao et al. 2019b)

With the production pressure fixed around 22.1 MPa, the injection pressure (bottom hole pressure at depth of -2450m) under different injection rates was measured at injection point and shown in Fig. 5.30. Because of viscosity difference, the injection pressure grows with continuous water injection, while the CO_2 injection induced pressure drops down dramatically at the beginning of production and then gradually stabilizes at a relative low level. Especially, there remains a minor decease of the CO_2 injection induced pressure after several years production, as the priority channel comes to production point. In general, the injection pressure increases with higher injection rate. This tendency is more significant in water injection due to its low compressibility and high viscosity. After 30 years of production, the pressure difference between injection and production point reaches 27.0 MPa and 4.4 MPa respectively for water and CO_2 injection with the rate of 100kg/s. This is also result from the significant difference of compressibility and viscosity between CO_2 and water. Fig. 5.31 shows the cumulative produced net energy, which is calculated according to Eq. 5.27 (Randolph & Saar 2011). The cumulative produced net energy increases with production time. However, its increasing rate gradually slows down because of decreasing production temperature. In spite of this, the accumulated net energy shows a positive relation with injection rate. And produced net energy of water injection is obviously higher than CO_2 injection under the same injection rate. Approximately, the net energy of 1.03×10^{16}J, 1.43×10^{16}J are produced for water injection with the injection rate of 60 kg/s, 80 kg/s respectively. This produced net energy is comparable as the one with CO_2 injection at the rates of 100kg/s and 120kg/s, respectively.

$$H = Q(h - h_0) \tag{5.31}$$

Where h is the enthalpy of produced fluid and h_0 is the fluid enthalpy at injection condition, Q is production rate.

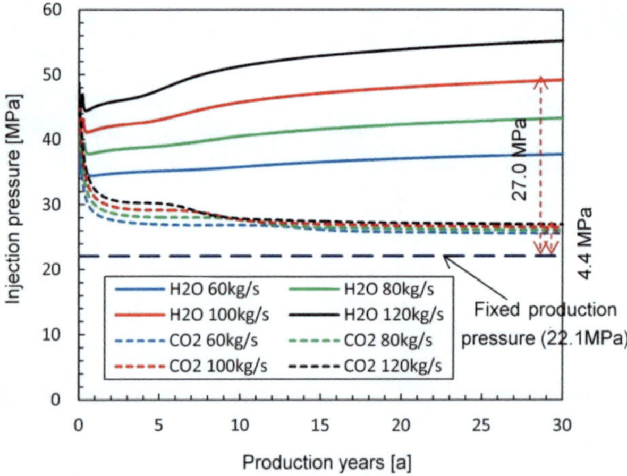

Figure 5.30. The injection pressure at different rates of different fluid (Liao et al. 2019b)

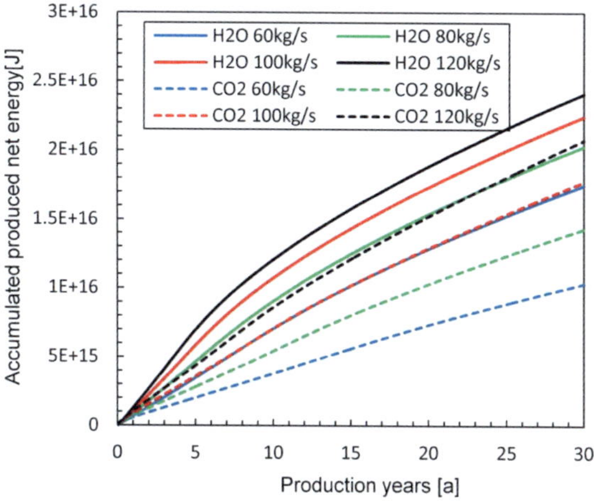

Figure 5.31. the accumulated produced net energy during production (Liao et al. 2019b)

After 30 years of production, the produced temperature and driving pressure (pressure difference between injection and production point) under different injection rates for CO_2 and water are potted in Fig. 5.32. It shows a positive relationship between driven pressure and injection rate, while a negative relationship between produced temperature and injection rate. In addition to this, the produced temperature for CO_2 injection is about 12 °C higher than water injection for the same injection rate. The driving pressure of water injection increases much faster with injection rate than that of CO_2 injection. However, the driving pressure due to CO_2 injection, which grows slightly from 3.9MPa to 4.9 MPa with rate from 60kg/s to 120kg/s, is negligible. However, the difference of driving pressure between water and CO_2 injection increases dramatically from 12.8 MPa to 28.2 MPa with injection rate from 60kg/s to 120kg/s. The average thermal capacity of different injection fluid over 30 years is plotted against different injection rates in Fig. 5.33. A general rule that the average thermal capacity increases with rising injection rate has been proven in both water and CO_2 injection. Meanwhile, the water injection owns a higher geothermal capacity. But with the water injection rate increases, the growing rate of average thermal capacity slows down. There are two reasons account for it. Firstly, the heat is relatively overexploited under higher water injection rate, which is confirmed by the produced temperature curve (see Fig. 5.27). Secondly, the fluid flow meets greater resistance under higher injection rate, which is confirmed by injection pressure (see Fig. 5.30). Such tendency is not clearly observed in CO_2 injection. Therefore, the average thermal capacity difference between water and CO_2 injection decreases gradually from 7.52 MWth to 3.54 MWth with injection rate from 60kg/s to 120kg/s. Specifically, the injection rate corresponding the average thermal capacity of 18.4 MWth for water injection is 60 kg/s, but it should be 100 kg/s for CO_2 to achieve the same thermal capacity. Based on these, it is recommended to use CO_2 as working fluid for heat extraction, for its low driving pressure and relative stable heat mining.

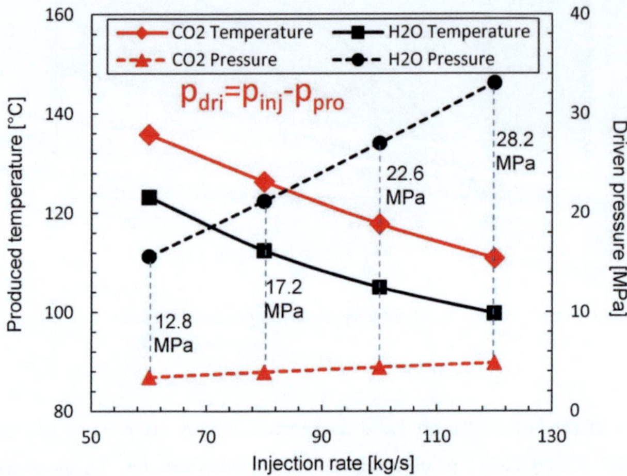

Figure 5.32. Comparing the produced temperature after 30a production and driving pressure (Liao et al. 2019b)

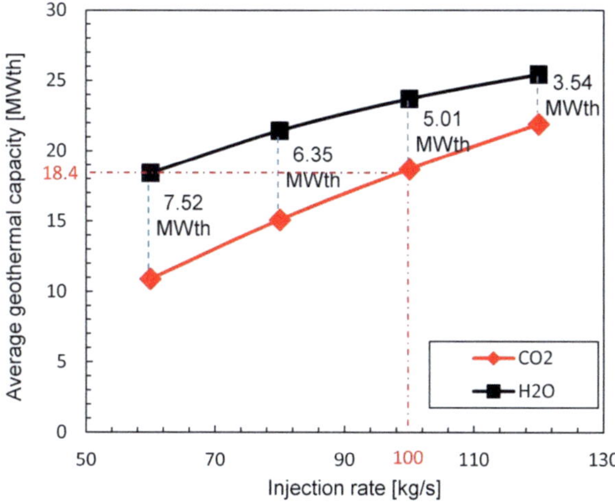

Figure 5.33. Comparing the averaged geothermal capacity at different injection rate (Liao et al. 2019b)

5.10. CO_2 sequestration in the lifetime of Dikili EGS

Due to pressure difference and friction, the injected CO_2 leaks into surrounding formation or is directly stored in the reservoir, which could be treated as CO_2 sequestration. In the lifetime of geothermal project, the CO_2 sequestration can be divided into 4 stages: reservoir stimulation, mass circulation, CO_2 circulation and CO_2 balance. As shown by the injection and production rate in Fig. 5.34, the CO_2 is injected at 100kg/s without production in stage I, which namely fracturing to create fracture network. Stage II is the phase for building mass circulation. In this phase, it takes about 4 months for mass circulation. Meanwhile, the CO_2 production rate growing gradually from 0 to almost 97 kg/s. Then the CO_2 production rate keeps for a constant for approximately 8 years, during which time the priority channel gradually moves to and eventually reaches the production point. This period is assigned as stage III namely CO_2 circulation. After that, it entries into CO_2 balance stage, in which the CO_2 production increases further close to but still lower than the injection rate of 100kg/s. As shown in Fig. 5.35, the sequestrated CO_2 is still accumulated at the loss rate in stage IV. The amounts of sequestrated CO_2 are 0.09, 0.24, 0.34 and 0.28 million tons in the stage from I to IV, respectively.

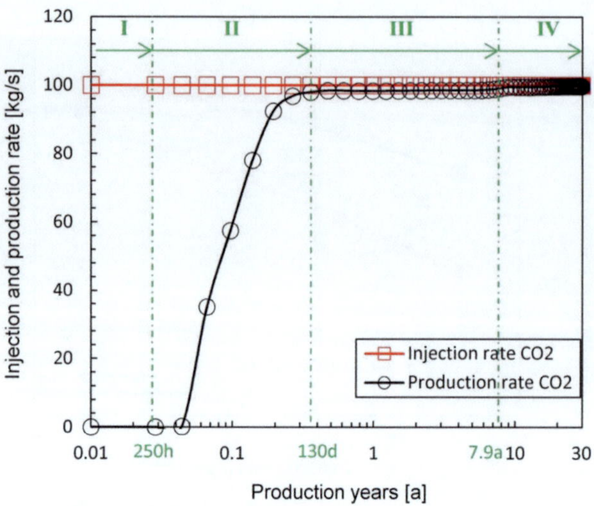

Figure 5.34. Comparing the injection rate and production rate in the case of CO_2 injection at 100 kg/s (Liao et al. 2019b)

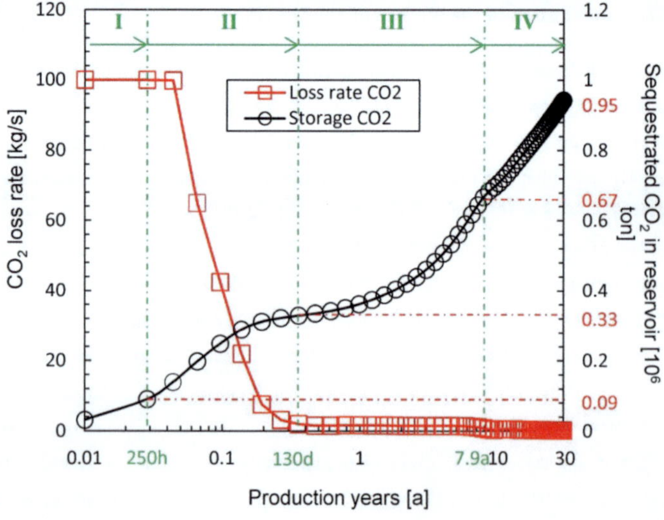

Figure 5.35. Illustrated the CO_2 sequestrated rate and the accumulated sequestrated content of CO_2 during production (Liao et al. 2019b)

The total amount of sequestrated CO_2 over 30 years production against different injection rate is plotted in Fig. 5.36. Obviously, there is a positive correlation between total sequestrated CO_2 and injection rate. The amount of sequestrated CO_2 grows from 0.73, 0.85, 0.95 to 1.02 million tons with CO_2 injection rate rising from 60, 80, 100 to 120 kg/s respectively, demonstrating the potential contribution of CO_2-ESG on the mitigation of the emission of CO_2.

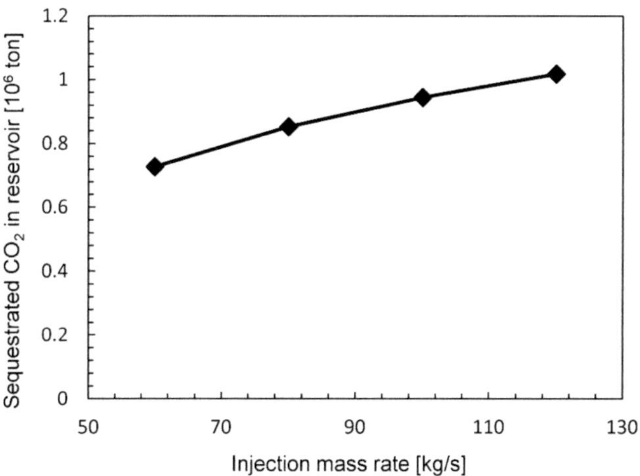

Figure 5.36. The sequestrated content of CO_2 versus injection rate in the lifetime of geothermal project (Liao et al. 2019b)

5.11. Summary

In this chapter, a 3D continuum anisotropic damage-permeability model is developed in the THM framework of TOUGH2MP (TMVOC)-FLAC3D. Based on this model, the stimulation, heat extraction in the Dikili geothermal reservoir with formation water is conducted by using CO_2 as an alternative working fluid. Besides, the ancillary benefit of CO_2 sequestration has been discussed in the lifetime of CO_2-EGS. Some distinctive conclusions are summarized as follows:

A 3D field scale approach based on continuum anisotropic damage-permeability model is proposed to study stimulation and heat extraction of a geothermal reservoir with pre-existing formation water by using supercritical CO_2.

In the application, a fracture network with maximum half-length of 1200m, a height of 1000 m and an average width of 37.8 m are created after injecting 90,000 kg CO_2 in 250 h. The maximum permeability of 1.3×10^{-13} m2 is achieved in x- and z-direction. Finally, a SRV of 8.71×10^7 m³ and SRA of 2.24×10^6 m2are yielded at 250 h.

A priority channel with high gas saturation will be formed during heat production because of the viscosity difference between CO_2 and water. The heat is mainly mined in this priority channel.

Generally, for both water and CO_2 injection, the driving pressure and average thermal capacity shows a positive correlation with injection rate, while the eventual produced temperature shows a negative correlation with injection rate. Under the same mass injection rate, the driving pressure and average thermal capacity of water injection is higher than CO_2 injection, whereas the eventual produced temperature of water injection is lower. Therefore, CO_2 as working fluid for heat extraction have the benefits of low driving pressure and is beneficial for realizing relative stable heat mining.

The injected CO_2 will be detained in geothermal reservoir due to leak off, which could be regarded as geologically sequestrated CO_2.The eventual sequestrated CO_2 has a positive correlation with injection rate. At the end of 30 years geothermal production, up to 950,000 tons CO_2 is sequestrated under the injection rate of 100kg/s, demonstrating the potential contribution of CO_2-ESG on the sequestration of CO_2.

6. Conclusion

Although the renewable energy production is increasing continuously, the geo-energy, including oil, gas as well as coal, still contributes the most towards world energy demand and consumptions. With the enhancement of public awareness, the eco-friendly technique in geo-energy exploitation has attracted more and more attention. In this thesis, two specific THM coupled models are developed for geothermal reservoir and unconventional gas reservoir, respectively, to study the performance of supercritical CO_2 in geo-energy exploitation.

After systematic literature study in chapter 2, it has been found that utilization of supercritical CO_2 for geothermal energy extraction from deep geothermal reservoirs is still a new concept. Currently, there is no implemented or planned CO_2-based geothermal project. In spite of this, it has attracted extensive attentions, as CO_2 re-utilization not only addresses the serious problems caused by traditional fracturing fluid, like ground water pollution, water resource shortage etc., but also partly mitigates the greenhouse emission through CO_2 geological sequestration. In comparison with geothermal reservoir, the CO_2 utilization in unconventional gas reservoirs started much earlier, where it has dominated the petroleum industry especially in the field of enhanced oil recovery. In addition to its advantages in geothermal reservoirs, further benefits of utilizing supercritical CO_2 are observed in unconventional gas reservoirs, such as production recovery desorption of CH_4 by CO_2, low break down pressure etc. However, the technology of CO_2 utilization in geo-energy production still faces many problems and challenges, such as the proppant transport by CO_2 in unconventional gas reservoir, geothermal reservoir stimulation by CO_2 etc.

As the properties of CO_2 are highly temperature and pressure sensitive and many complex mechanical responses are involved during energy production, the new models should be integrated under a THM coupled framework. Therefore, the fundamentals of used THM framework (TOUGH2MP-FLAC3D) are introduced in chapter 3. Equation of state (EOS), TMVOC, is adopted for it is possible to define additional components by the user. In this multi-flow simulator, maximum three phases, including gas phase, aqueous phase and hydrate phase can be solved simultaneously. The governing equations and numerical algorithm for mechanic, hydraulic and thermal processes have been introduced in detail. Moreover, the coupling between different individual processes and data flow between two simulators has been discussed as well.

Attempting to study the performance of supercritical CO_2 in unconventional gas reservoir, a specific numerical model able to characterize fracture propagation as well as leak-off is developed in chapter 4. In this model, a pre-defined potential fracture plane is inserted in the middle of host reservoir elements, perpendicular to minimum stress. On this potential fracture plane, different specific models, including fracture model, proppant transport model and multi flow model, are developed and integrated in the popular THM framework TOUGH2MP-FLAC3D. Then, the feasibility of these models has been verified separately by three simple cases. Besides, different fracturing fluids (mainly based upon temperature-dependent viscosity) are defined firstly, including pure CO_2, two thickened CO_2s and guar gum. Other properties, like density and enthalpy, are assumed equal to the base fluid of fracturing fluid. After that, by using different fracturing fluids (pure CO_2, two thickened CO_2s, pure water and guar gum), the fracturing has been carried out on a fictitious model with the properties of a typical tight gas reservoir. Their performances in fracture creation are compared on the base of simulated results. Additionally, since the properties of CO_2 are closely correlated to temperature, the thermal effect has been compared under different thermal conditions. Finally, the proppants with different densities are used to address the weak proppant-carrying ability of CO_2. Some important conclusions are drawn from these simulation results. Firstly, it is found that pure CO_2 is not efficient to create a fracture in such tight gas reservoir. However, the fracturing ability of CO_2 can be improved significantly, through adding CO_2 thickener, for example surfactant. In comparison with traditional fracturing fluid guar gel, the leak-off of thickened CO_2 is significantly more. But considering the expansion of CO_2 in the fracture, the fracking ability of thickened CO_2 becomes comparable with traditional guar gel fluid. According to the simulation results, the fracking ability can be improved by about 37% due to CO_2 expansion. Secondly, a linear correlation is observed between the break down pressure and fracturing fluid viscosity. The break down pressure rises with increasing fracturing fluid viscosity. Thirdly, comparing fracture width profiles under different thermal conditions discovers that the fracture width profile is determined by the combination of CO_2 expansion and leak-off rate. In non-isothermal condition, the temporal and spatial leak-off rate and expansion effect are unevenly distributed. Due to higher temperature, larger leak-off rate and expansion effect are found around fracture tip. Fourthly, at the beginning of fracturing, the fluid expansion plays a critical role in fracture creation. With continuous injection, the dominating factor in fracture creation is gradually switched from expansion to leak-off. Overall, leak-off has more significant influence on fracture-creation than fluid expansion in the whole process. Lastly, in the study of proppant transport, thickened CO_2 with light proppant can achieve a better proppant placement than heavy proppant, even better as the one transported by guar gum. Due to low specific heat capacity, thickened CO_2 has a quick gel breaking speed, resulting in better fracture support.

In order to study the critical engineering processes, reservoir stimulation and heat extraction in deeper geothermal reservoirs, a 3D simulator based on a continuum anisotropic model and anisotropic permeability model has been developed under the THM framework of TOUGH2MP-FLAC3D (TMVOC). In geothermal energy exploitation, the stress state change due to fracturing fluid injection can induce a directional damage in geothermal reservoir, which further triggers an anisotropic permeability. At the macro level, this process is same as fracture propagation. In addition, the impact of effective stress is considered in this model as well. It is believed that usually a fracture network is stimulated in the geothermal reservoir with brittle rocks, particularly volcanic rock. This damage model and permeability is verified by triaxial test in laboratory and cyclic fracturing test in flied. The plausibility of this model is also validated by a series of case studies on a simple 2D model. On a real case study from Dikili geothermal project, the reservoir stimulation and heat extraction in the presence of formation water with different production rates by using supercritical CO_2 is conducted and compared with the results by using water-based fluid. Besides, the ancillary benefit of CO_2 sequestration has been discussed in the lifetime of CO_2-EGS under the effect of different injection rates. Finally, some distinctive conclusions are drawn from this simulation. A useful 3D field scale approach based on continuum anisotropic damage-permeability model is proposed which has the ability to study stimulation and heat extraction of a deep geothermal reservoir with pre-existing formation water by using supercritical CO_2. In the application at Dikili project, a fracture network with maximum half-length of about 1200m, a height of 1000 m and an average width of 37 m are created after injecting 90,000 kg CO_2 in 250 h. The maximum permeability of $1.3 \times 10^{-13}\,m^2$ is achieved in x- and z-direction. The final SRV of $8.7 \times 10^7\,m^3$ and SRA of $2.2 \times 10^6\,m^2$ are yielded at 250 h. During heat production, a priority channel with high gas saturation is formed because of the viscosity difference between CO_2 and water. The heat is mainly mined in this priority channel. Generally, for both water and CO_2 injection, the driving pressure and average thermal capacity shows a positive relation with injection rate, while the ultimate produced heat shows inverse relationship with injection rate. Under the same mass injection rate, the driving pressure and average thermal capacity of water injection is higher than CO_2 injection, whereas the eventual produced heat of water injection is lower. Therefore, CO_2 as working fluid for heat extraction has the advantage of low driving pressure and is beneficial for realizing relatively stable heat mining. Moreover, the injected CO_2 detained in geothermal reservoir due to leak off can be regarded as geologically sequestrated CO_2. The amount of sequestrated CO_2 has a positive relation with injection rate. At the end of 30 years geothermal production, up to 950,000 tons CO_2 is sequestrated at an injection rate of 100kg/s, demonstrating the potential contribution of CO_2-EGS on the sequestration of CO_2.

7. References

Alber, M., Solibida, C. (2017). Geomechanical characterization of a host rock for enhanced geothermal system in the North-German Basin. *Procedia engineering*, 191, 158-163.

Al Hinai, N. M., Saeedi, A., Wood, C. D., Myers, M., Valdez, R., Sooud, A. K., Sari, A. (2018). Experimental evaluations of polymeric solubility and thickeners for supercritical CO_2 at high temperatures for enhanced oil recovery. *Energy & fuels*, 32(2), 1600-1611.

Altunin, V. V. (1975) Thermophysical Properties of Carbon Dioxide. Publishing House of Standards, Moscow, USSR (in Russian) pp:551.

Al-Busaidi, A., Hazzard, J. F., Young, R. P. (2005). Distinct element modeling of hydraulically fractured Lac du Bonnet granite. *Journal of geophysical research: solid earth*, 110(B6).

Armstead, H. C. H., Tester, J. W. (1987). Heat mining: a new source of energy, 29. West 35th Street, New York NY, 10001.

Asanuma, H., Tsuchiya, N., Muraoka, H., Ito, H. (2015). Japan beyond-brittle project: development of EGS beyond brittle-ductile transition. In *Proceedings World Geothermal Congres*, Melbourne, Australia.

Atrens, A. D., Gurgenci, H., Rudolph, V. (2008). CO_2 thermosiphon for competitive geothermal power generation. *Energy & fuels*, 23(1), 553-557.

Barree, R. D., Conway, M. W. (1994). Experimental and numerical modeling of convective proppant transport. In *SPE Annual Technical Conference and Exhibition*. Society of Petroleum Engineers.

Bayraktar, Z., Kiran, E. (2000). Miscibility, phase separation, and volumetric properties in solutions of poly (dimethylsiloxane) in supercritical carbon dioxide. *Journal of applied polymer science*, 75(11), 1397-1403.

BGR. (2019). GeneSys Project Aktuelles. Retrieved November, 12, 2019, from https://www.genesys-hannover.de/Genesys/DE/Aktuelles/aktuelles_node.html

Borgia, A., Pruess, K., Kneafsey, T. J., Oldenburg, C. M., Pan, L. (2012). Numerical simulation of salt precipitation in the fractures of a CO_2-enhanced geothermal system. *Geothermics*, 44, 13-22.

Brown, D. W. (2000). A hot dry rock geothermal energy concept utilizing supercritical CO_2 instead of water. In *Proceedings of the twenty-fifth workshop on geothermal reservoir engineering*, Stanford University (pp. 233-238).

Brooks, R. H., Corey, A. T. (1964). Hydraulic properties of porous media, hydrology papers, no. 3, colorado state university, ft. *Collins, Colo.*

Chen, Y., Hu, S., Zhou, C., Jing, L. (2014). Micromechanical modeling of anisotropic damage-induced permeability variation in crystalline rocks. *Rock mechanics and rock engineering, 47*(5), 1775-1791.

Chen, L., Wang, C. P., Liu, J. F., Liu, J., Wang, J., Jia, Y., Shao, J. F. (2015). Damage and plastic deformation modeling of Beishan granite under compressive stress conditions. *Rock Mechanics and Rock Engineering, 48*(4), 1623-1633.

Chiarelli, A. S., Shao, J. F., Hoteit, N. (2003). Modeling of elastoplastic damage behavior of a claystone. *International Journal of plasticity, 19*(1), 23-45.

Cooper, H. W., Simmons, G. (1977). The effect of cracks on the thermal expansion of rocks. *Earth and Planetary Science Letters, 36*(3), 404-412.

Cooper, J. R., Dooley, R. B. (1994). IAPWS release on surface tension of ordinary water substance. *International Association for the Properties of Water and Steam (IAPWS)*, Charlotte, NC, 2.

Cui, G., Ren, S., Zhang, L., Wang, Y., Zhang, P. (2018). Injection of supercritical CO_2 for geothermal exploitation from single-and dual-continuum reservoirs: Heat mining performance and salt precipitation effect. *Geothermics, 73*, 48-59.

Damjanac, B., Cundall, P. (2016). Application of distinct element methods to simulation of hydraulic fracturing in naturally fractured reservoirs. *Computers and Geotechnics, 71*, 283-294.

Davies, J. P., Davies, D. K. (1999). Stress-dependent permeability: characterization and modeling. In *SPE Annual Technical Conference and Exhibition*. Society of Petroleum Engineers.

Dobson, P., Kneafsey, T. J., Blankenship, D., Valladao, C., Morris, J., Knox, H., ... & Mattson, E. (2017). An introduction to the EGS Collab Project. *GRC Transactions, 41*, 837-849.

Doherty, M. D., Lee, J. J., Dhuwe, A., O'Brien, M. J., Perry, R. J., Beckman, E. J., Enick, R. M. (2016). Small molecule cyclic amide and urea based thickeners for organic and sc-CO_2/organic solutions. *Energy & Fuels, 30*(7), 5601-5610.

Dontsov, E. V. (2017). An approximate solution for a plane strain hydraulic fracture that accounts for fracture toughness, fluid viscosity, and leak-off. *International Journal of Fracture, 205*(2), 221-237.

Economides, M., Nolte, K. 2000. Reservoir Stimulation, 3rd edition. Chichester, UK: John Wiley & Sons.

Enick, R. M., Olsen, D. K., Ammer, J. R., Schuller, W. (2012). Mobility and Conformance Control for CO_2 EOR via Thickeners, Foams, and Gels--A Literature Review of 40 Years of Research and Pilot Tests. In *SPE improved oil recovery symposium*. Society of Petroleum Engineers.

Etheridge, D. M., Steele, L. P., Langenfelds, R. L., Francey, R. J., Barnola, J. M., Morgan, V. I. (2001). Law dome atmospheric CO_2 data. *Data Contribution Series*, 83. LAW DOME

Fairless, C. M., Joseph, W. (1986). Effective Well Stimulations With Gelled Methanol/Carbon Dioxide Fracturing Fluids. In *SPE East Texas Regional Meeting*. Society of Petroleum Engineers.

Falta, R. W., Pruess, K., Finsterle, S., Battistelli, A. (1995). T2VOC user's guide.

Fang, C., Chen, W., Amro, M. (2014). Simulation study of hydraulic fracturing using super critical CO_2 in shale. In *Abu Dhabi International Petroleum Exhibition and Conference*. Society of Petroleum Engineers.

Feng, Y., Chen, X., Xu, X. F. (2014). Current status and potentials of enhanced geothermal system in China: a review. *Renewable and Sustainable Energy Reviews*, 33, 214-223.

Frank, E. D., Sullivan, J. L., Wang, M. Q. (2012). Life cycle analysis of geothermal power generation with supercritical carbon dioxide. *Environmental Research Letters*, 7(3), 034030.

Gandossi, L. (2013). An overview of hydraulic fracturing and other formation stimulation technologies for shale gas production. Luxembourg: European Commision - Joint Research Centre, Institute for Energy and Transport.

Gaurav, A., Dao, E. K., Mohanty, K. K. (2012). Evaluation of ultra-light-weight proppants for shale fracturing. *Journal of Petroleum Science and Engineering*, 92, 82-88.

Gou, Y., Hou, Z., Li, M., Feng, W. Liu, H. (2016). Coupled thermo–hydro–mechanical simulation of CO_2 enhanced gas recovery with an extended equation of state module for TOUGH2MP-FLAC3D. Journal of Rock Mechanics and Geotechnical Engineering, 8(6), 904-920.

Godec, M. L., Kuuskraa, V. A., Dipietro, P. (2013). Opportunities for using anthropogenic CO_2 for enhanced oil recovery and CO_2 storage. *Energy & Fuels*, 27(8), 4183-4189.

Gou, Y. (2018) Numerical study of coupled THMC processes related to geo-energy production. (Doctoral dissertation, Causthal University of Technology).

Grigoli, F., Cesca, S., Rinaldi, A. P., Manconi, A., Lopez-Comino, J. A., Clinton, J. F., ... & Wiemer, S. (2018). The November 2017 Mw 5.5 Pohang earthquake: A possible case of induced seismicity in South Korea. *Science, 360*(6392), 1003-1006.

Gunter, W. D., Mavor, M. J., Robinson, J. R. (2005). CO_2 storage and enhanced methane production: field testing at Fenn-Big Valley, Alberta, Canada, with application. In *Greenhouse Gas Control Technologies* 7 (pp. 413-421). Elsevier Science Ltd.

Halm, D., Dragon, A. (1996). A model of anisotropic damage by mesocrack growth; unilateral effect. *International Journal of Damage Mechanics, 5*(4), 384-402.

Heimlich, C., Masson, F., Schmittbuhl, J., Ferhat, G. (2016). Geodetic measurements for geothermal site monitoring at Soultz-sous-Forêts and Rittershoffen deep geothermal sites. In *Processing European Geothermal Congress*(Vol. 125).

Hill, B., Hovorka, S., Melzer, S. (2013). Geologic carbon storage through enhanced oil recovery. Energy Procedia, 37, 6808-6830.

Holm, L. W. (1986). Miscibility and miscible displacement. *Journal of Petroleum Technology, 38*(08), 817-818.

Hong, L., Thies, M. C., Enick, R. M. (2005). Global phase behavior for CO_2-philic solids: the CO_2+ β-d-maltose octaacetate system. The Journal of supercritical fluids, 34(1), 11-16.

Hou, M. Z., Wundram, L., Wendel, H., Meyer, R., Schmidt, M., Kretzschmar, H. J., Kretzschmar, S., Schmitz, O., Franz, O. (2011). Developing, modeling and in-situ-testing a long-term wellbore seal in the framework of the CO_2-EGR research project CLEAN in the natural gas field Altmark. *Energy Procedia*, 4, 5291-5298

Hou, Z., Gou, Y., Taron, J., Gorke, U. J., Kolditz, O. (2012). Thermo-hydro-mechanical modeling of carbon dioxide injection for enhanced gas-recovery (CO_2-EGR): a benchmarking study for code comparison. *Environmental Earth Sciences*, 67(2), 549-561.

Hou, M. Z., Zhou, L., Weichmann, M., Li, M., Feng, W. (2015a) A New Numerical 3D-model for Simulation of The Whole Hydraulic Fracturing Process in Consideration of Hydro-mechanical Coupling Effects. *DGMK/ÖGEW Frühjahrstagung.* 22.-23.04. 2015 in Celle

Hou Z, Şen O, Gou Y, Eker AM, Li M, Yal GP, Cambazoğlu S, Were P (2015b) Preliminary geological, geochemical and numerical study on the first EGS project in Turkey. *Environmental Earth Sciences,* 73(11):6747-6767.

IPCC. (2018). Summary for Policymakers of IPCC Special Report on Global Warming of 1.5°C approved by governments. Retrieved November, 10, 2019, from https://www.ipcc.ch/2018/10/08/summary-for-policymakers-of-ipcc-special-report-on-global-warming-of-1-5c-approved-by-governments/

Irvine, T.F. Jr and Liley, P.E. (1984). Steam and Gas Tables with Computer Equations. *Academic Press Inc.*, Orlando, Florida, 185 pp.

Ishida, T., Aoyagi, K., Niwa, T., Chen, Y., Murata, S., Chen, Q., Nakayama, Y. (2012). Acoustic emission monitoring of hydraulic fracturing laboratory experiment with supercritical and liquid CO_2. *Geophysical Research Letters*, 39(16).

Itasca (2011) Description of Formulation with Validation Problems Revision 1.0. ITASCA Consulting Group, Inc

Jeanne, P., Rutqvist, J., Rinaldi, A. P., Dobson, P. F., Walters, M., Hartline, C., Garcia, J. (2015). Seismic and aseismic deformations and impact on reservoir permeability: The case of EGS stimulation at The Geysers, California, USA. Journal of Geophysical Research: *Solid Earth*, 120(11), 7863-7882.

Jiang, T., Shao, J. F., Xu, W. Y., & Zhou, C. B. (2010). Experimental investigation and micromechanical analysis of damage and permeability variation in brittle rocks. *International Journal of Rock Mechanics and Mining Sciences*, 47(5), 703-713.

Jiang, J., Shao, Y., Younis, R. M. (2014). Development of a multi-continuum multi-component model for enhanced gas recovery and CO_2 storage in fractured shale gas reservoirs. In *SPE improved oil recovery symposium*. Society of Petroleum Engineers.

Kraft, T., Mai, P. M., Wiemer, S., Deichmann, N., Ripperger, J., Kästli, P., Bachmann, C. E., Fäh, D., Wossner, J., Giardini, D. (2009). Enhanced Geothermal Systems in Urban Areas-Lessons Learned from the 2006 Basel ML3. 4 Earthquake. In *AGU Fall Meeting Abstracts*.

Kuuskraa, V. A., Koperna, G. J. (2006). Evaluating the potential for 'game changer' improvements in oil recovery efficiency from CO2 enhanced oil recovery. *Prepared for US Department of Energy, Office of Fossil Energy—Office of Oil and Natural Gas*.

Lee, S. H., & Ghassemi, A. (2009). Thermo-poroelastic finite element analysis of rock deformation and damage. In *43rd US Rock Mechanics Symposium & 4th US-Canada Rock Mechanics Symposium*. American Rock Mechanics Association.

Lee, I. H., Ni, C. F. (2015). Fracture-based modeling of complex flow and CO_2 migration in three-dimensional fractured rocks. *Computers & geosciences*, 81, 64-77.

Lee, J. (2017). Small Molecule Associative CO_2 Thickeners for Improved Mobility Control (Doctoral dissertation, University of Pittsburgh).

Li, M., Hou, M. Z., Zhou, L., Gou, Y. (2018). Numerical study of the hydraulic fracturing and energy production of a geothermal well in Northern Germany. In *Geomechanics and Geodynamics of Rock Masses: Selected Papers from the 2018 European Rock Mechanics Symposium* (p. 415). CRC Press.

Li, S., Zhang, D. (2019). How effective is carbon dioxide as an alternative fracturing fluid?. *SPE Journal*, *24*(02), 857-876.

Li, S., Feng, X. T., Zhang, D., Tang, H. (2019a). Coupled thermo-hydro-mechanical analysis of stimulation and production for fractured geothermal reservoirs. *Applied Energy*, 247, 40-59.

Li, Q., Wang, Y., Wang, F., Li, Q., Kobina, F., Bai, H., Yuan, L. (2019b). Effect of a modified silicone as a thickener on rheology of liquid CO_2 and its fracturing capacity. *Polymers*, 11(3), 540.

Lillies, A. T., King, S. R. (1982). Sand fracturing with liquid carbon dioxide. In *SPE Production Technology Symposium*. Society of Petroleum Engineers.

Liao, J., Hou, M. Z., Mehmood, F., Feng, W. (2019a). A 3D approach to study the interaction between hydraulic and natural fracture. *Environmental Earth Sciences*, *78*(24), 681.

Liao, J., Cao, C., Hou, Z., Mehmood, F., Feng, W., Yue, Y., Liu, H., (2019b). Field scale numerical modeling of heat extraction in geothermal reservoir based on fracture network creation with supercritical CO_2 as working fluid. *Geoenviromental earth science*. Under review.

Liao, J., Gou, Y., Feng, W., Mehmood, F., Xie, Y., Hou, Z. (2020a). Development of a full 3D numerical model to investigate the hydraulic fracture propagation under the impact of orthogonal natural fractures. *Acta Geotechnica*, *15*(2), 279-295.

Liao, J., Hou, Z., Haris, M., Tao, Y., Xie, Y., Yue, Y. (2020b). Numerical evaluation of hot dry rock reservoir through stimulation and heat extraction using a three-dimensional anisotropic coupled THM model. *Geothermics*, *83*, 101729.

Liu, Y. (2006). Settling and hydrodynamic retardation of proppants in hydraulic fractures. (Doctoral dissertation, University of Texas at Austin).

Liu, H. J. (2014). Numerical study of physicochemical interactions for CO_2 sequestration and geothermal energy utilization in the Ordos Basin, China. (Doctoral dissertation, Causthal University of Technology).

Liu, Y., Wang, G., Yue, G., Lu, C., Zhu, X. (2017). Impact of CO_2 injection rate on heat extraction at the HDR geothermal field of Zhacanggou, China. *Environmental Earth Sciences*, 76(6), 256.

Liu, L., Zhu, W., Wei, C., Elsworth, D., Wang, J. (2018). Microcrack-based geomechanical modeling of rock-gas interaction during supercritical CO_2 fracturing. *Journal of Petroleum Science and Engineering*, 164, 91-102.

Lu, S. M. (2018). A global review of enhanced geothermal system (EGS). *Renewable and Sustainable Energy Reviews*, 81, 2902-2921.

Luo, F., Xu, R. N., Jiang, P. X. (2014). Numerical investigation of fluid flow and heat transfer in a doublet enhanced geothermal system with CO_2 as the working fluid (CO_2–EGS). *Energy*, 64, 307-322.

Luo, X., Wang, S., Wang, Z., Jing, Z., Lv, M., Zhai, Z., Han, T. (2015). Experimental investigation on rheological properties and friction performance of thickened CO_2 fracturing fluid. *Journal of Petroleum Science and Engineering*, 133, 410-420.

Lyons, W. (2010) Working Guide to Petroleum and Natural Gas Production Engineering. Gulf Professional Publishing. ISBN:9781856178457

Ma, T., Chen, P., Zhao, J. (2016). Overview on vertical and directional drilling technologies for the exploration and exploitation of deep petroleum resources. *Geomechanics and Geophysics for Geo-Energy and Geo-Resources*, 2(4), 365-395.

Magliocco, M., Kneafsey, T. J., Pruess, K., & Glaser, S. (2011). Laboratory experimental study of heat extraction from porous media by means of CO2. In *36th Workshop on Geothermal Reservoir Engineering*.

Maleki, K., Pouya, A. (2010). Numerical simulation of damage–Permeability relationship in brittle geomaterials. *Computers and Geotechnics*, 37(5), 619-628.

Maurer, V., Cuenot, N., Gaucher, E., Grunberg, M., Vergne, J., Wodling, H., Lehujeur, M., Schmittbuhl, J. (2015). Seismic monitoring of the Rittershoffen EGS project (Alsace, France). In *Proceedings World Geothermal Congress* (pp. 19-25).

Middleton, R. S., Carey, J. W., Currier, R. P., Hyman, J. D., Kang, Q., Karra, S., Jiménez-Martínez, J., Porter, M. L., Viswanathan, H. S. (2015). Shale gas and non-aqueous fracturing fluids: Opportunities and challenges for supercritical CO_2. *Applied Energy*, 147, 500-509.

Mills, T., Humphreys, B. (2013). Habanero pilot project—Australia's first EGS power plant. In *Proceedings* (pp. 37-41).

Muller, N., Qi, R., Mackie, E., Pruess, K., Blunt, M. J. (2009). CO_2 injection impairment due to halite precipitation. *Energy procedia*, 1(1), 3507-3514.

Narasimhan, T. N., Witherspoon, P. A. (1976). An integrated finite difference method for analyzing fluid flow in porous media. *Water Resources Research*, 12(1), 57-64.

NOAA ESRL (2019). Trends in Atmospheric Carbon Dioxide, Retrieved November, 05, 2019, from ftp://ftp.cmdl.noaa.gov/ccg/co2/trends/co2_annmean_mlo.txt

Nuckols, E. B., Miles, D., Laney, R., Polk, G., Friddle, H., Simpson, G., Baroid, N. L. (1981). Drilling fluids and lost circulation in hot dry rock geothermal wells at Fenton Hill (No. LA-UR-80-3604; CONF-810105-1). Los Alamos Scientific Lab., NM (USA).

Olasolo, P., Juárez, M. C., Morales, M. P., Liarte, I. A. (2016). Enhanced geothermal systems (EGS): A review. *Renewable and Sustainable Energy Reviews*, 56, 133-144.

Olson, J. E. (2004). Predicting fracture swarms—The influence of subcritical crack growth and the crack-tip process zone on joint spacing in rock. *Geological Society, London, Special Publications*, 231(1), 73-88.

Olson, J. E., Taleghani, A. D. (2009). Modeling simultaneous growth of multiple hydraulic fractures and their interaction with natural fractures. In *SPE hydraulic fracturing technology conference*. Society of Petroleum Engineers.

Oldenburg, C. M. (2003). Carbon sequestration in natural gas reservoirs: Enhanced gas recovery. In *Natural Gas Storage", Proceedings of TOUGH Symposium 2003, LBNL*.

Oldenburg, C. M., Doughty, C., Spycher, N. (2013). The role of CO_2 in CH_4 exsolution from deep brine: Implications for geologic carbon sequestration. *Greenhouse Gases: Science and Technology*, 3(5), 359-377.

Paik, I. H., Tapriyal, D., Enick, R. M., Hamilton, A. D. (2007). Fiber Formation by Highly CO_2-Soluble Bisureas Containing Peracetylated Carbohydrate Groups. *Angewandte Chemie International Edition*, 46(18), 3284-3287.

Pan, C., Chávez, O., Romero, C. E., Levy, E. K., Corona, A. A., Rubio-Maya, C. (2016). Heat mining assessment for geothermal reservoirs in Mexico using supercritical CO_2 injection. *Energy*, 102, 148-160.

Parker, J. C., Lenhard, R. J., Kuppusamy, T. (1987). A parametric model for constitutive properties governing multiphase flow in porous media. *Water Resources Research*, 23(4), 618-624.

Peng, D. Y., Robinson, D. B. (1976). A new two-constant equation of state. *Industrial & Engineering Chemistry Fundamentals*, *15*(1), 59-64.

Peng, P., Ju, Y., Wang, Y., Wang, S., Gao, F. (2017). Numerical analysis of the effect of natural microcracks on the supercritical CO_2 fracturing crack network of shale rock based on bonded particle models. *International Journal for Numerical and Analytical Methods in Geomechanics*, 41(18), 1992-2013.

Perera, M., Gamage, R., Rathnaweera, T., Ranathunga, A., Koay, A., Choi, X. (2016). A review of CO_2-enhanced oil recovery with a simulated sensitivity analysis. *Energies*, 9(7), 481.

Pruess, K., Battistelli, A. (2002). TMVOC, a numerical simulator for three-phase non-isothermal flows of multicomponent hydrocarbon mixtures in saturated-unsaturated heterogeneous media.

Pruess, K., Garcia, J. (2002). Multiphase flow dynamics during CO_2 disposal into saline aquifers. *Environmental Geology*, 42(2-3), 282-295.

Pruess, K. (2004). The TOUGH codes—A family of simulation tools for multiphase flow and transport processes in permeable media. *Vadose Zone Journal*, 3(3), 738-746.

Pruess, K. (2006). Enhanced geothermal systems (EGS) using CO_2 as working fluid—A novel approach for generating renewable energy with simultaneous sequestration of carbon. *Geothermics*, 35(4), 351-367.

Pruess, K., Azaroual, M. (2006). On the feasibility of using supercritical CO_2 as heat transmission fluid in an engineered hot dry rock geothermal system. In *Proceedings, Thirty-First Workshop on Geothermal Reservoir Engineering*. Citeseer.

Pruess, K. (2007). Enhanced Geothermal Systems (EGS) comparing water with CO_2 as heat transmission fluids.

Randolph, J. B., Saar, M. O. (2010). Coupling geothermal energy capture with carbon dioxide sequestration in naturally permeable, porous geologic formations: A comparison with enhanced geothermal systems. *Geothermal Research Council Transactions*, 34, 433-438.

Randolph, J. B., Saar, M. O. (2011). Impact of reservoir permeability on the choice of subsurface geothermal heat exchange fluid: CO_2 versus water and native brine. In *Proceedings for the geothermal resources council 35th annual meeting*, (pp. 23-26).

Reid, R. C., Prausnitz, J. M., Poling, B. E. (1987). The properties of gases and liquids. McGraw-Hill, New York.

7. References

Rutqvist, J., Tsang, C. F. (2002). A study of caprock hydromechanical changes associated with CO_2-injection into a brine formation. *Environmental Geology*, 42(2-3), 296-305.

Rutqvist, J., Wu, Y. S., Tsang, C. F., Bodvarsson, G. (2002). A modeling approach for analysis of coupled multiphase fluid flow, heat transfer, and deformation in fractured porous rock. *International Journal of Rock Mechanics and Mining Sciences*, 39(4), 429-442.

Sanaei, A., Shakiba, M., Varavei, A., Sepehrnoori, K. (2018). Mechanistic modeling of clay swelling in hydraulic-fractures network. *SPE Reservoir Evaluation & Engineering*, 21(01), 96-108.

SDS Energy (2014) Dikili EGS field informative report

Schill E, Genter A, Cuenot N, Kohl T (2017) Hydraulic performance history at the Soultz EGS reservoirs from stimulation and long-term circulation tests. *Geothermics* 70:110-124.

Shi, C., Huang, Z., Beckman, E. J., Enick, R. M., Kim, S. Y., Curran, D. P. (2001). Semi-fluorinated trialkyltin fluorides and fluorinated telechelic ionomers as viscosity-enhancing agents for carbon dioxide. *Industrial & engineering chemistry research*, 40(3), 908-913.

Shi, Y., Song, X., Shen, Z., Wang, G., Li, X., Zheng, R.,Geng, L., Li, J., Zhang, S. (2018). Numerical investigation on heat extraction performance of a CO_2 enhanced geothermal system with multilateral wells. *Energy*, 163, 38-51.

Smith, M. M., Walsh, S. C., McNab, W. W., Carroll, S. A. (2013). Experimental investigation of brine-CO_2 flow through a natural fracture: permeability increases with concurrent dissolution/reprecipitation reactions (No. LLNL-CONF-615372). Lawrence Livermore National Lab.(LLNL), Livermore, CA (United States).

Souley, M., Homand, F., Pepa, S., Hoxha, D. (2001). Damage-induced permeability changes in granite: a case example at the URL in Canada. *International Journal of Rock Mechanics and Mining Sciences*, 38(2), 297-310.

Soave, G. (1972). Equilibrium constants from a modified Redlich-Kwong equation of state. *Chemical engineering science*, 27(6), 1197-1203.

Sun, Z. X., Zhang, X., Xu, Y., Yao, J., Wang, H. X., Lv, S., Huang, Y., Cai, M., Huang, X. (2017). Numerical simulation of the heat extraction in EGS with thermal-hydraulic-mechanical coupling method based on discrete fractures model. *Energy*, 120, 20-33.

Tapriyal, D. (2009). Design of non-fluorous CO_2 soluble compounds. (Doctoral dissertation, University of Pittsburgh).

Torres, M. D., Hallmark, B., Wilson, D. I. (2014). Effect of concentration on shear and extensional rheology of guar gum solutions. *Food Hydrocolloids*, 40, 85-95.

Trickett, K., Xing, D., Enick, R., Eastoe, J., Hollamby, M. J., Mutch, K. J., Rogers S. E., Heenan, R. K., Steytler, D. C. (2009). Rod-like micelles thicken CO_2. *Langmuir*, 26(1), 83-88.

Ueda, A., Kato, K., Ohsumi, T., Yajima, T., Ito, H., Kaieda, H., Metcalfe, H., Takase, H. (2005). Experimental studies of CO_2-rock interaction at elevated temperatures under hydrothermal conditions. *Geochemical Journal*, 39(5), 417-425.

Van der Waals, J. D. (1873). On the continuity of the gas and liquid state (Doctoral dissertation, University of Leiden, Leiden, The Netherlands).

Van Genuchten, M. T. (1980). A closed-form equation for predicting the hydraulic conductivity of unsaturated soils 1. *Soil science society of America journal*, *44*(5), 892-898.

Vargaftik, NB. (1975) Tables on the Thermophysical Properties of Liquids and Gases, second ed. John Wiley & Sons, New York, NY, USA pp.758.

Vengosh, A., Jackson, R. B., Warner, N., Darrah, T. H., Kondash, A. (2014). A critical review of the risks to water resources from unconventional shale gas development and hydraulic fracturing in the United States. *Environmental science & technology*, 48(15), 8334-8348.

Wan, Y., Xu, T., Pruess, K. (2011). Impact of fluid-rock interactions on enhanced geothermal systems with CO_2 as heat transmission fluid. In *Thirty-Sixth Workshop on Geothermal Reservoir Engineering Stanford University*.

Wang, Y., Hong, L., Tapriyal, D., Kim, I. C., Paik, I. H., Crosthwaite, J. M., Hanilton A. D., Thies, M. C., Beckman, E. J., Enick R. M., Johnson, J. K. (2009). Design and evaluation of nonfluorous CO_2-soluble oligomers and polymers. *The Journal of Physical Chemistry B*, 113(45), 14971-14980.

Wang, C., Winterfeld, P. H., Wu, Y. S., Wang, Y., Chen, D., Yin, C., Pan, Z. (2014). Coupling Hydraulic Fracturing Propagation and Gas Well Performance for Simulation of Production in Unconventional Shale Gas Reservoirs. In *AGU Fall Meeting Abstracts*.

Wang, H. (2015). Numerical modeling of non-planar hydraulic fracture propagation in brittle and ductile rocks using XFEM with cohesive zone method. *Journal of Petroleum Science and Engineering*, 135, 127-140.

Wang, J., Elsworth, D., Wu, Y., Liu, J., Zhu, W., Liu, Y. (2018). The influence of fracturing fluids on fracturing processes: a comparison between water, oil and SC-CO_2. *Rock Mechanics and Rock Engineering*, 51(1), 299-313.

Ward, H., Eykelbosh, A., Nicol, A. M. (2016). Addressing uncertainty in public health risks due to hydraulic fracturing. *Environmental Health Review*, 59(2), 57-61.

Watanabe, N., Egawa, M., Sakaguchi, K., Ishibashi, T., Tsuchiya, N. (2017). Hydraulic fracturing and permeability enhancement in granite from subcritical/brittle to supercritical/ductile conditions. *Geophysical Research Letters*, 44(11), 5468-5475.

Wei, Z., Zhang, D. (2013). A fully coupled multiphase multicomponent flow and geomechanics model for enhanced coalbed-methane recovery and CO_2 storage. *SPE Journal*, 18(03), 448-467.

Wu, H., Kitagawa, A., Tsukagoshi, H. (2005). Development of a portable pneumatic power source using phase transition at the triple point. In *Proceedings of the JFPS International Symposium on Fluid Power* (Vol. 2005, No. 6, pp. 310-315). The Japan Fluid Power System Society.

Wu, K., Olson, J. E. (2013). Investigation of the impact of fracture spacing and fluid properties for interfering simultaneously or sequentially generated hydraulic fractures. *SPE Production & Operations*, 28(04), 427-436.

Wu, K., Olson, J. E. (2015). Simultaneous multifracture treatments: fully coupled fluid flow and fracture mechanics for horizontal wells. *SPE journal*, 20(02), 337-346.

Wu, X., Zhang, Y., Sun, X., Huang, Y., Dai, C., Zhao, M. (2018). A novel CO_2 and pressure responsive viscoelastic surfactant fluid for fracturing. *Fuel*, 229, 79-87.

Xu, T., Pruess, K. (2010). Reactive transport modeling to study fluid-rock interactions in enhanced geothermal systems (EGS) with CO_2 as working fluid. In *Conference Paper Presented at the World Geothermal Congress*, Bali, Indonesia (pp. 25-29).

Xu, T., Feng, G., Hou, Z., Tian, H., Shi, Y., Lei, H. (2015). Wellbore–reservoir coupled simulation to study thermal and fluid processes in a CO_2-based geothermal system: identifying favorable and unfavorable conditions in comparison with water. *Environmental earth sciences*, 73(11), 6797-6813.

Xu, R., Li, R., Ma, J., He, D., Jiang, P. (2017). Effect of mineral dissolution/precipitation and CO2 exsolution on CO_2 transport in geological carbon storage. *Accounts of chemical research*, 50(9), 2056-2066.

Yang, Y., Chen, H., Wang, W. (2013). Implementation and numerical verfication of Elastoplastic-anisotropic damgae consitutive model in FLAC3D. *Journal of Yangtze River Scientific Research Institute* 30(12):48-53.

Yan, C., Zheng, H., Sun, G., Ge, X. (2016). Combined finite-discrete element method for simulation of hydraulic fracturing. *Rock Mechanics and Rock Engineering*, 49(4), 1389-1410.

Yin, S., Dusseault, M. B., Rothenburg, L. (2011). Coupled THMC modeling of CO_2 injection by finite element methods. *Journal of Petroleum Science and Engineering*, 80(1), 53-60.

Yoon, J. S., Zimmermann, G., Zang, A. (2015). Numerical investigation on stress shadowing in fluid injection-induced fracture propagation in naturally fractured geothermal reservoirs. *Rock Mechanics and Rock Engineering*, 48(4), 1439-1454.

Yost, A. B., Mazza, R. L., Gehr, J. B. (1993). CO_2/Sand fracturing in devonian shales. In *SPE Eastern Regional Meeting*. Society of Petroleum Engineers.

Zang, A., Stephansson, O., Stenberg, L., Plenkers, K., Specht, S., Milkereit, C., Schill, E., Dresen, G., Zimmermann, G., Dahm, T., Weber, M. (2016). Hydraulic fracture monitoring in hard rock at 410 m depth with an advanced fluid-injection protocol and extensive sensor array. *Geophysical Journal International*, 208(2), 790-813.

Zhao, Q., Lisjak, A., Mahabadi, O., Liu, Q., Grasselli, G. (2014). Numerical simulation of hydraulic fracturing and associated microseismicity using finite-discrete element method. *Journal of Rock Mechanics and Geotechnical Engineering*, 6(6), 574-581.

Zhang, L. M., Kong, T., Hui, P. S. (2007). Semi-dilute solutions of hydroxypropyl guar gum: Viscosity behaviour and thixotropic properties. *Journal of the Science of Food and Agriculture*, 87(4), 684-688.

Zhang, S., She, Y., Gu, Y. (2011). Evaluation of polymers as direct thickeners for CO_2 enhanced oil recovery. *Journal of Chemical & Engineering Data*, 56(4), 1069-1079.

Zhang, C. L. (2016). The stress–strain–permeability behaviour of clay rock during damage and recompaction. *Journal of Rock Mechanics and Geotechnical Engineering*, 8(1), 16-26.

Zhang, X., Lu, Y., Tang, J., Zhou, Z., Liao, Y. (2017). Experimental study on fracture initiation and propagation in shale using supercritical carbon dioxide fracturing. *Fuel*, 190, 370-378.

Zhang, N., Wei, M., Bai, B. (2018). Statistical and analytical review of worldwide CO_2 immiscible field applications. *Fuel*, 220, 89-100.

Zhang, Q., Ma, D., Liu, J., Wang, J., Li, X., Zhou, Z. (2019). Numerical simulations of fracture propagation in jointed shale reservoirs under CO_2 fracturing. *Geofluids*, 2019.

Zhou, L. (2014). New numerical approaches to model hydraulic fracturing in tight reservoirs with consideration of hydro-mechanical coupling effects. (Doctoral dissertation, Causthal University of Technology).

Zhou, L., Gou, Y., Hou, Z., Were, P. (2015). Numerical modeling and investigation of fracture propagation with arbitrary orientation through fluid injection in tight gas reservoirs with combined XFEM and FVM. *Environmental earth sciences*, 73(10), 5801-5813.

Zhou, C., Remoroza, A. I., Shah, K., Doroodchi, E., Moghtaderi, B. (2016). Experimental study of static and dynamic interactions between supercritical CO_2/water and Australian granites. *Geothermics*, 64, 246-261.

Zimmermann, G., Zang, A., Stephansson, O., Klee, G., Semiková, H. (2019). Permeability Enhancement and Fracture Development of Hydraulic In Situ Experiments in the Äspö Hard Rock Laboratory, Sweden. *Rock Mechanics and Rock Engineering*, 52(2), 495-515.

Resume

Last Name: Liao

Fist Name: Jianxing

Brith: 1991.12.10

Nationality: Chinese

2010.09-2012.08	Bachelor study of Civil Engineering in Sichuan University, China
2013.03-2014.09	Bachelor study of Geoenvironmental Engineering in TU Clausthal, Germany
2014.10-2016.09	Master study of Geoenvironmental Engineering in TU Clausthal, Germany
2016.10-2020.06	Doctoral study of Engineering in TU Clausthal, Germany

Publication during PhD studies:

Liao, J., Hou, M. Z., Mehmood, F., & Feng, W. (2019). A 3D approach to study the interaction between hydraulic and natural fracture. Environmental Earth Sciences, 78(24), 681.

Liao, J., Gou, Y., Feng, W., Mehmood, F., Xie, Y., & Hou, Z. (2020). Development of a full 3D numerical model to investigate the hydraulic fracture propagation under the impact of orthogonal natural fractures. Acta Geotechnica, 15(2), 279-295.

Liao, J., Hou, Z., Haris, M., Tao, Y., Xie, Y., & Yue, Y. (2020). Numerical evaluation of hot dry rock reservoir through stimulation and heat extraction using a three-dimensional anisotropic coupled THM model. Geothermics, 83, 101729.

Liao, J., Cao, C., Hou, Z., Mehmood, F., Feng, W., Yue, Y., & Liu, H. (2020). Field scale numerical modeling of heat extraction in geothermal reservoir based on fracture network creation with supercritical CO_2 as working fluid. Environmental Earth Sciences, 79, 1-22.